PENGUIN BOOKS
THE NARMADA DAMMED

A one-time software professional, Dilip D'Souza now writes for a living. His first book was *Branded by Law: Looking at India's Denotified Tribes*. This is his second book. He lives in Bombay with his wife and son.

W0043163

The Narmada Dammed

An Inquiry into the Politics of Development

Dilip D'Souza

PENGUIN BOOKS

An imprint of Penguin Random House

PENGUIN BOOKS

USA | Canada | UK | Ireland | Australia
New Zealand | India | South Africa | China | Singapore

Penguin Books is part of the Penguin Random House group of companies
whose addresses can be found at global.penguinrandomhouse.com

Published by Penguin Random House India Pvt. Ltd
4th Floor, Capital Tower 1, MG Road,
Gurugram 122 002, Haryana, India

Penguin
Random House
India

First published by Penguin Books India 2002

Copyright © Dilip D'Souza 2002

All rights reserved

10 9 8 7 6 5 4 3 2

The views and opinions expressed in this book are the authors' own and the
facts are as reported by them which have been verified to the extent possible,
and the publishers are not in any way liable for the same.

ISBN 9780143028659

Typeset in Nebraska by SÜRYA, New Delhi
Printed at Repro India Limited

This book is sold subject to the condition that it shall not, by way of trade
or otherwise, be lent, resold, hired out, or otherwise circulated without the
publisher's prior consent in any form of binding or cover other than that in
which it is published and without a similar condition including this condition
being imposed on the subsequent purchaser.

www.penguin.co.in

MIX
Paper from
responsible sources
FSC® C047271

This is a legitimate digitally printed version of the book and therefore might not
have certain extra finishing on the cover.

To my parents, who know the road less-travelled.

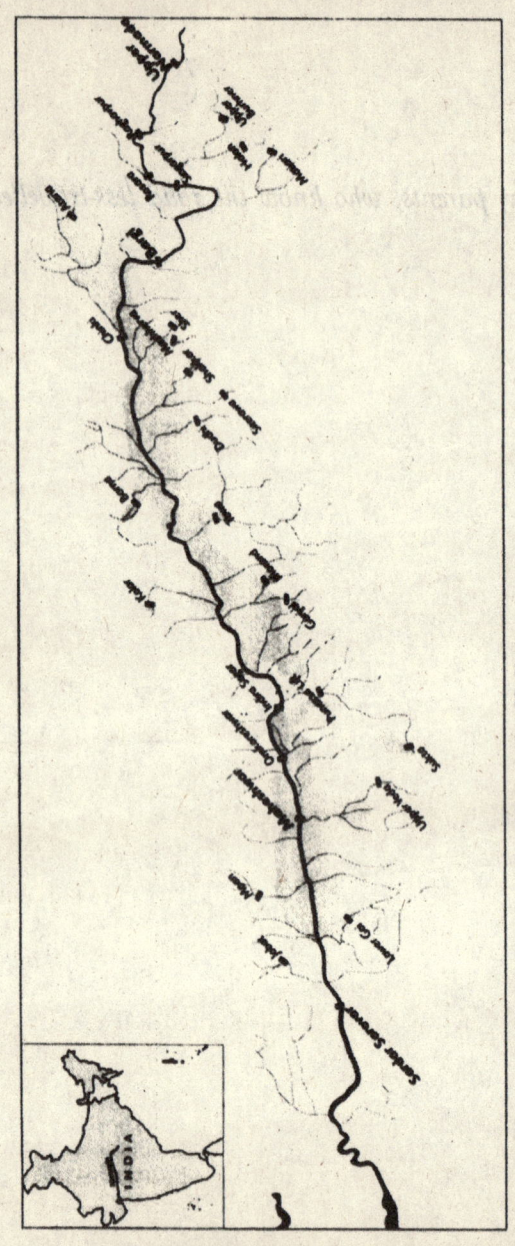

Map of Proposed Big Dams in the Narmada Valley

Contents

Acknowledgements

Various people have read parts of this book, as it appeared in different forms in different places. Their comments and our discussions persuaded me that this book was possible. As always, they are too numerous to name, but I think they know who they are. Thank you.

Someone once said I needed a 'ruthless' editor. I don't know if Kamini Mahadevan was ruthless, but she has certainly been meticulous and thorough. I'll owe you one, Kamini.

Jacques Leslie, Amma, Bain and Vibha read my chapters, offered thoughts that made me think, and helped better the book. Doff of the hat to you all.

And, of course, I learned from innumerable people I met on my travels. Their experiences, spirit, knowledge, also gave me things to think about, to write about. Whether on the banks of the river or in far-away Kutch, the Narmada is their story; it speaks of their dreams and lives. They are the soul of this book.

Moments of Epiphany

Truly, one of the defining moments of my life came late one wet and gusty night in August 1996. I was on a path that led to a little town called Bargi Nagar in Madhya Pradesh. At least, I think I was on a path. It was hard to be sure. It was so dark that I could find my way only by following sounds of the footsteps of the man in front of me. Five of us had been tramping for hours in incessant rain through soggy, muddy fields, stumbling over boulders and fighting through thorny hedges. By the time somebody said we had found the 'path', we were hungry, exhausted and longing most of all for a cup of *chai*.

Then came that moment. Without warning, someone coughed softly, almost in my face. That set a dog barking right at my feet. I walked on, but it took the best part of the next thirty minutes for my nerves to stop twanging violently.

What had happened was this: as we trudged towards Bargi Nagar, we passed a tiny settlement, walking right past—or maybe through, who knows?—a house there. In the humid darkness, I had not even seen the house. I only realized it was there when its owner coughed. That's how dark it was.

But as we walked, we could see bright lights off in the

distance. Tube lights that lit the streets and *dhabas* of Bargi Nagar. Multicoloured lights that bathed the temple on the hill above the town. Floodlights that showed off the enormous Bargi dam on the Narmada river, some distance beyond the town.

It was those lights that defined this moment for me. As I walked, looking at them but wrapped in night, through a village wrapped in night, a lot of things snapped into place.

Here was an enormous dam that supplied electricity not just to Bargi Nagar, but further afield too. (I am conscious, even as I write these words in Bombay, that the electricity that lets me work comes from some far-off dam. One just like Bargi.) Yet the authorities who built and operate Bargi dam had not thought it necessary to provide electricity to the tiny village we had just unwittingly visited.

This tiny village, which remains as dark as it has for centuries.

Yes, a great deal became very clear to me that night: the meaning of all arguments about dams, about the very idea of 'development' as we have practised it in India. When that man coughed, it all made sense.

This sense: The way India has developed has left just too many Indians in the dark.

There are many reasons I think it is a mistake to build dams, especially large ones; there are many books and papers that will explain these reasons. But if I manage to put together figures and arguments to make my case against these large dams—the dams on the Narmada in particular—those who built them and want to build some more can just as easily gather evidence to support their case just as well. To an otherwise indifferent observer, I suppose it boils down to whose set of arguments seems more appealing. Besides, after a point, who cares? If the years of debate over the Sardar Sarovar dam have shown those of us with an interest in the issue anything, it is that most people are just sick of the numbers, the polemics, the rationales, the tension, that substitute for debate. I get sick myself.

Nevertheless, being sick doesn't mean the questions go away.

If a dam produces electricity that entirely bypasses people who live right there next to the dam; if these very people must make great sacrifices so such dams can be built; if this happens over and over again in an independent and democratic country—then there is something profoundly wrong about building these dams. There is something fundamentally warped in how we have chosen to 'develop'. And yet, for over half a century now, this is really how we have pursued development. In ways that have left many Indians in darkness.

I wrote this book to examine and understand this 'development'. To make the case that even if you think dams are a good idea, you should be concerned at the way they have been built in India. At what they have done, what development itself has done, to Indians, to India.

•

Let me expand on that theme by telling you about another moment of epiphany.

It's the evening rush hour on the last day of 1995. Loaded with several bags, two women and I struggle into a sardine-can of a second-class compartment at Bombay Central. Trying to keep our balance as we talk, we chug north on the train. Just before Bandra, we stop, as suburban trains sometimes do, for ten inexplicable minutes.

Ordinarily, these strange halts are annoying and frustrating. But this time, I am positively grateful. Because it means I can finish the interview I am conducting with one of these women, for a magazine article. This train ride from one meeting in town to another in the suburb of Goregaon is the only time Medha Patkar can give me on this particular trip to Bombay.

As crowds rush in and out and around us, as I fight to stay on my feet, to read my notes so I know what to ask next, to hold my little tape-recorder near enough to her face to catch her answers, I can't help a stray thought. How much easier it would be to do this over strong coffee and a vada-sambar at Kamat's restaurant! And yet that thought is replaced very quickly by another. In many ways, this situation—interviewing Medha Patkar in a train crammed with Bombay commuters—is entirely appropriate. In many ways, the very processes this woman has fought for years have resulted in these brutal commuting conditions for so many million Bombayites.

Bombay is certainly a magnet for much of India. Why is that so? Because if you are stuck in a tiny village in Madhya Pradesh, let's say, with no job, no schools for your children, no health-care facilities for miles around, barely any income for sustaining your family because the bit of land you till but don't even own doesn't bring in enough— if you are in a place like that, Bombay begins to look very attractive indeed. In every sense of the word, there's opportunity here.

So for years, people have come to this city in search of opportunity. Because the way we have developed has neglected our rural areas. Besides, even when 'development' has touched rural areas, it has uprooted people and sent them scurrying to the cities. I mean development like dams, power plants, test firing ranges, oil refineries and uranium processing factories: all these markers of progress have invariably displaced people from their homes. Many of them end up in our cities.

That's one reason Bombay has so many people today, and its trains are so impossibly crowded, even outside the rush hour. Inexplicably, our progress has ignored the way the majority of the country lives and travels and works. Ever wondered why the same money, time, debate and effort that has gone into building flyovers for Bombay has not gone into making bus travel here—or rail travel, for that matter—a more pleasant experience? After all, buses carry several times more road commuters than cars do. You'd think the effort would be to ease conditions for this majority: but buses don't use the flyovers because stops can't be situated on them.

Over the years, Medha Patkar and the Narmada Bachao Andolan (NBA) have fought much more than just the construction of a dam on the Narmada river. They have fought to question this very idea of development. What is it? What has it done for us? Whose development is it? Whom has it benefited? Whom has it hurt?

●

One last defining moment. This one came when I read a letter I have a copy of in my papers. Dated 2 December 1992, it's from the commissioner in the Government of India's Ministry of Water Resources, N. Suryanarayanan, on the Ministry's letterhead, and addressed to Medha Patkar. In this letter, Mr Suryanarayanan says:

> The Sardar Sarovar Project as envisaged now will reduce substantially the distress due to drought in Kutch, Saurashtra and North Gujarat *by say 2025 AD*. [Emphasis added]

Consider the meaning of that casual 'by say 2025 AD'. The letter was written well before the NBA filed its case in the Supreme Court, stopping construction on the dam for five years. It was written, that is, well before the NBA was able to cause any delay at all in the progress of the project. And at that time, this government official announced that the dam—this 'lifeline of Kutch and Saurashtra', those drought-parched districts of Gujarat—would relieve drought distress there 'by say 2025 AD'. A good thirty-three years, a full generation, in the future. A date so remote as to be effectively meaningless.

When I read this letter, I sat back and thought: these men talk so easily of this dam being the 'only hope' of Kutch and Saurashtra. Well, the good folks in those districts hope for water now, today. They don't hope for a promise that's to be met—if at all—at some essentially indefinite time in the future. What must they feel to hear that their thirst for water will be slaked thirty-three years from now?

I imagine they must feel frustrated, even somewhat shortchanged, that while no particular urgency attaches to addressing their distress *today*, that very distress is pronounced as the reason for building this dam.

•

This is the context in which I want to place this book.

After all, as their champions will tirelessly point out, dams do bring benefits. Electricity, water, improved agricultural production, the control of floods, and so on. In India, we have grown up learning about the great boon of Punjab's Green Revolution, brought about not just by Norman Borlaug and his radically new strains of grain, but also by the water the Bhakra-Nangal dam delivers across the state. And you can find more such dams, such gifts to India, all over the country—Hirakud, Tungabhadra, Nagarjunasagar and many more.

Besides, it was by building all these dams—and we Indians have been some of the world's most industrious dam-builders—that free India showed the world it was a nation that could stand on its own. I suspect there's a good argument to be made that in the first decades after independence, these dams gave us the self-respect and worth that a young nation seeks. Hadn't Jawaharlal Nehru reminded us, on that ecstatic day in 1947, of our 'tryst with destiny'? And hadn't Nehru himself famously proclaimed dams the 'temples of modern India'? 'What a stupendous, magnificent work,' said Nehru as he walked around the Bhakra dam site, 'a work which only that nation can take up which has faith and boldness!' (McCully: 1)

And, in fact, it is that spirit that suffuses the building of the dams on the Narmada. I have a 1988 leaflet about the Sardar Sarovar dam, which lists four 'Unique features of the project':

1. About 7 million [cubic metres] of chilled concrete will be placed in the Dam. This is the highest quantity of concrete to be placed in any dam in India. This quantity is equivalent to the quantity that would be normally required to construct say one lane concrete road around the equator.

2. . . . For erecting [cable ways to lay concrete] height of A-frame required will be nearly 1½ times the height of the famous Kutub Minar. Such a huge system involving placement of concrete at a rate of 7000 cubic metres/day has never been operated in India.

3. The Underground River Bed Power House . . . will be one of the biggest hydro electric underground power house in the country.

4. The main canal . . . will be largest capacity Irrigation Canal in the World. (Sardar Sarovar Narmada Project leaflet).

Note the tone: 'highest', 'biggest', 'largest', 'never been operated'—the idea is to inspire awe, to imply achievement on a vast scale. And it works. It's certainly in that grand tradition of dams: awe-inspiring, imposing, vast structures that prove Indian technology is second to none.

Why, then, the questions today about dams?

Because there was a darkness behind these shimmering visions. Over the years, the people our dams displaced had, without exception, been treated in a manner that brought shame to the dreams and ideals of independent India. They were summarily shoved off land they had called their own for generations. Left to fend for themselves as best they could. Left to watch as that land disappeared under the long lakes that ballooned out behind the new dams.

This is no mere accusation. In a publicity booklet about the Sardar Sarovar dam, the then-chairman of the Sardar Sarovar Narmada Nigam Limited (SSNNL), Sanat Mehta, says: '[T]his is the *first ever* project of the country wherein rehabilitation problem was considered in much details [sic]. . . . [If] the project would have been executed in the old fashion, its cost would have been far less.' [Emphasis added] (Mehta, no date).

Mehta goes on to praise the rehabilitation package for the Narmada oustees, but these sentences say a lot. Through the first few decades of independence, rehabilitation was never a priority with our dam builders. That was the way it was in the 'old fashion' style of dam building.

To explain this away, the popular wisdom was that it was a question of trade-offs. If you wanted progress, 'some' people would have to sacrifice. After all, it was in the national interest.

A fine sounding mantra, and it kept the shame successfully hidden for many years. But it couldn't last.

Soon, some Indians began to ask increasingly uncomfortable questions. Who are these people who are expected to sacrifice? How many are there, have there been? Why are the same people affected so often, some of them twice and thrice over? Is this trade-off itself reasonable? Is progress always to be achieved at their cost, and if so, is it truly progress? Is it truly in the national interest? In fact, whose is the national interest, really?

As plans for damming the Narmada took shape through the 1960s, 1970s and 1980s, there was no reason to expect anything but the same story to unfold: of the state's indifference to the issue of rehabilitation, despite people's questions. And yet it was these very questions that resulted in the plans for resettlement and rehabilitation (R&R) that Sanat Mehta praises in that booklet.

To me, all of this says that it is time that we began discussing dams not just in terms of their costs and benefits. Certainly there is a value to that kind of analysis and it must be done, but, as we have seen on the Narmada, the debate always ends in stalemate. To me, a definitely interested observer, it has become increasingly clear that dams are not built because of, or after, a careful consideration of costs and benefits. They are, above all, symbols. In our early years, they were symbols of Indian prowess. More recently, and especially in Gujarat, they have become symbols of a particular notion of progress, of that elusive idea of 'development.'

Why are such symbols important? There is an unstated, though hardly subtle, logic at work here. After all, surely every true Indian must want his country to progress and develop. And if dams are designated markers of that progress, we must build them, mustn't we? The builders of dams want us to make just this link in our minds. And those who speak out against dams, why, they must want

India to remain backward and undeveloped. What's more, such people cannot be true Indians. In fact, they are anti-national.

Development is thus easily conflated with a concept such as patriotism—because this conflation allows opponents to be branded as anti-national, their arguments ridiculed for that reason rather than on the merits. This is a theme that runs through whatever debate takes place about dams today. In fact, in Gujarat, criticizing the Sardar Sarovar dam is seen as akin to urging the secession of Kashmir. Not something to be done lightly, and certainly not something that any political party would indulge in. Which is why every political party in the state has made so much political capital on the Sardar Sarovar dam. In the end, dams are political symbols. That's why they are built. That's the way we must debate them.

•

In *Cadillac Desert*, a marvellous exploration of water policy and dam-building in the USA, Marc Reisner describes just what I am alluding to here. He writes about a grandiose plan that was proposed in the 1960s in Texas, that true home of the grandiose. These planners wanted to divert the Mississippi river from below New Orleans—a city that is essentially at the mouth of the river. They would send the water over 1,200 miles of marshlands, swamps and plains, climbing—climbing!—through 3,600 feet. All this, to bring relief to the residents of west Texas, who were tapping their groundwater as if it was limitless. A fantastic chimera of a plan. But Reisner tells us:

> There followed ... one of those peculiar metamorphoses in which a plan, as it evolves, conforms less and less to constraints of nature, economics and thermodynamics and more and more to the stridency

of certain constituents and the desires of certain elected officials. (Reisner: 461)

The Texas Water Plan, as it was called, eventually got shown up for what it was. Being the plan it was, a political one above all, it died the death it deserved: it was defeated in a statewide referendum.

But the lesson remains, and I see no reason it cannot apply to us. It's time we understood dam-building as a political process. Doing so would, I believe, truly put our ideas of development in perspective.

•

Finally, a point about who this book is trying to reach, and how that drove the way I wrote it.

I began to think about it in the wake of the Supreme Court judgement of October 2000. Even if the Court dismissed the Narmada Bachao Andolan (NBA) petition against Sardar Sarovar, or maybe because it did, I thought it remained vital to consider the issues involved. As I indicated above, I want to appeal to the very people who believe that building big dams is a good thing; to show them that even so they should be concerned about how India has 'developed'. These dams may have brought them water and electricity, but the uneven development they lead to affects lives in profound ways. More than that, such development poses a threat in another sense: a society that ignores large numbers—leaves them in the dark—is asking for trouble.

I wanted to approach this task as an ordinary, if interested, citizen: to find out what I could about this project by reading the project authorities' own publicity material and other published literature, besides meeting other ordinary citizens like me who are interested in it. I did not want to speak to exalted officials, certainly no

elected ones, nor a collection of learned scholars. I just
wanted to see if the information that's generally available
to every one of us might say something about the dams on
the Narmada. Could that information by itself confirm my
beliefs, help build my case about the dams?

Besides, I know from personal experience that for
many people, the names 'Narmada Bachao Andolan' and
'Medha Patkar' are like red flags. Given how polarized the
debate over the dams on the Narmada has become, I can
understand that feeling. That is why I have also deliberately
avoided speaking to Patkar or other NBA activists for this
book. I have deliberately avoided defending them, or their
arguments. Enough other books make that defence. I
believe that even had there been no NBA, the case I want
to make stands its ground. Please read this book in that
spirit.

After all, I don't want those red flags waving about
even before you begin reading it.

1

Dams as Temples

When dams are discussed in India, you'll hear one term a lot: 'temple'. We owe that to Jawaharlal Nehru, who told us that they were the 'temples of modern India.' Inspired by his vision, we began building dams, and we rapidly became among the world's most diligent builders of dams.

There were two reasons for this. One, of course, was that we needed the electricity, the flood control, the irrigation and the drinking water. All the good things dams bring, or so we believed. The second reason was more subtle, but no less important. As I mentioned before, these dams were fine symbols of ambitious, hopeful, self-sufficient and finally independent India.

And so we built them. 1,137 large and major dams, according to one estimate. (McCully: 3) Their names have become part of our consciousness today: Damodar Valley, Tungabhadra, Hirakud, Bhakra-Nangal, Srisailam. In *Silenced Rivers*, Patrick McCully estimates that from 1947 to 1980, we spent about 15 per cent of our total national expenditure on building those thousand-plus dams (McCully: 20).

Given our enormous commitment to dams, it is startling that the Narmada, cutting through the middle of the country, had never been dammed. That is, till the late 1980s, when the Bargi dam, south of Jabalpur, was completed.

The Narmada, India's fifth-longest river at 1,312 km, empties into the Gulf of Khambat in Gujarat. For many years, planners had eyed this beautiful river. Instead of letting that water 'go waste' into the sea, they wanted to find a way, they told us, to bring its liquid bounty to parched, drought-stricken areas of Gujarat.

That, of course, meant building a dam, or a series of dams, on the river.

Thus was established the Narmada Valley Project: a plan to erect over 3,000 dams on the Narmada and its tributaries. The Sardar Sarovar Project (SSP) is one.

And what were these dams, the SSP in particular, going to do?

The simple answer is the one you hear most often: Sardar Sarovar will send water to thirsty Kutch and Saurashtra. Rainfall in Gujarat decreases as you travel west and north. The major urban centres—Ahmedabad, Baroda, Surat, Mehsana—generally get adequate rainfall, comparable to the average in other parts of the country. But Saurashtra gets far less rainfall; Kutch is even worse off. Vast stretches of Kutch are essentially desert—which is why it's called the Rann (literally, 'desert'). Every year, many residents of these areas are forced to migrate in search of water.

How do you address this problem? Well, what if the water in the Narmada, flowing out to sea through southern Gujarat, could find its way to Kutch? This was how the idea of damming the Narmada took root. At least on paper, at least as originally dreamed of, the Sardar Sarovar dam was not going to be built for the power it might supply, or the floods it might help control. Those might be side-benefits, but the real reason for the dam was that it would send water to some of the driest areas of India. Simple.

This quickly became such an assumed truth that today nobody questions the words used most often to describe the dam: It is the 'lifeline' of Kutch and Saurashtra, and more generally of Gujarat itself.

And yet, going by the plans of the builders themselves, Sardar Sarovar won't be solving the water problems of Kutch and Saurashtra any time soon. Other areas are far higher on the priority list for its water. The Gujarat government has already made much of how water from the Sardar Sarovar dam, after construction resumed in late 2000, is being supplied to Ahmedabad. Not thirsty Kutch. Here's how Ashish Kothari describes the project:

> The Sardar Sarovar Project is not about providing water to the thirsty lakhs in Kutch and Saurashtra, as its proponents have been arguing for decades. It is not about providing a lifeline to the drought-hit regions of Gujarat. It is more about facilitating the unending thirst for water and electricity of the big farmers, the industries, and the cities of central Gujarat ... more about repeating a failed model of 'development'. (Kothari)

Put another way, the decision to build the Sardar Sarovar dam was, at least in part, a political one. It was driven by the politically powerful constituencies of central Gujarat. For that reason, it was convenient, even necessary, to portray it as the 'lifeline' of a state. That was one way to sidestep criticism of the project. Being a lifeline, people who were sceptical about it were easily labelled ecological terrorists, or worse, anti-nationals, oblivious to the well-being of Kutch or even India. How easy to paint this scepticism as callousness towards the people of Kutch.

•

The idea of dams on the Narmada goes back to the famines that British India suffered from in the early twentieth century. Because producing more food was a major concern, the administration formed an Irrigation Commission to assess the irrigation potential of the country's major rivers. In 1904, it submitted a report to the British Parliament. It is filled with comments like:

> [T]here is a question of utilizing the waters of the Nerbudda [sic] ... [Its waters] could only in Gujarat be diverted to irrigate black soil areas which are to a very large extent quite unsuitable for irrigation. (Paranjpye: A-10)

> It is a very doubtful experiment to undertake irrigation works in a tract where its utility cannot be justified by existing agricultural practice, and I would deprecate any government measures to this end in the black soil of the Nerbudda valley. (Paranjpye: A-13)

Quoting these observations, the economist Vijay Paranjpye then writes:

> [T]he Commission felt that only the light alluvial soils lying at the foot of the Satpura ranges were capable of withstanding sustained irrigation. But the major portion of the river valley consists of deep black soils under which lie impervious strata. Hence, continued flood-irrigation would harm the soils and agriculture in the region. (Paranjpye: A-18)

Perhaps because of this turn-of-that-century assessment of what irrigation would do to agriculture, nothing more happened for several decades. But in 1946, the governments of Bombay and of the Central Provinces commissioned studies for developing the entire Narmada basin. After a study in 1947, the Central Waterways, Irrigation and Navigation Commission (CWINC) recommended seven sites to assess in greater detail.

The site where the Sardar Sarovar dam stands (Navagam) was first recommended in 1957, when an official of the Central Water and Power Commission (CWPC, the new name for the CWINC) visited the area. He realized that the rock formations in the river bed here would be useful while building the dam.

The CWPC prepared a firm proposal for the project in 1959 and gave it to the Government of Bombay (the province would be divided the following year into Gujarat and Maharashtra). The dam the CWPC proposed at Navagam was to be built in two stages, first to 160 feet and later to 300 feet. On the assumption that a 'high level canal' would be needed to send water to Kutch and Saurashtra, the government raised the proposed height to 320 feet.

After the two new states were born, Gujarat took over responsibility for the SSP. Prime Minister Jawaharlal Nehru laid the foundation stone for it on 5 April 1961.

That dam at Navagam has been the focus of controversy for years, but we should remember that it is only one of several planned for the river. As we will see, in 1965 the Khosla Committee recommended thirteen projects.

Why then did it take another three decades, and, in fact, over forty years after independence, for a dam to appear on this river? (The dam at Bargi).

A major reason is that the Narmada is unabashedly disrespectful of the lines we puny humans draw on maps. The river runs through three states—Madhya Pradesh,

Maharashtra and Gujarat—and for part of its length actually defines the borders between these states. Thus damming the river was a contentious proposition from the very start. The three states had to work out some kind of agreement over the costs and benefits of building such a dam. Who would pay what, who would take what?

Answering these questions took a while, and negotiations on the agreement went through fits and starts. There was no satisfactory mechanism for such intricate dialogue between states. The first effort to hammer out an agreement was made in 1965. Dr A.N. Khosla, who, as Chairman of the old CWINC, had examined and acted on the recommendations of the 1947 CWINC study, now headed a Narmada Water Resources Development Committee which spent a year drawing up a master plan for the use of the Narmada waters.

The Khosla Committee recommended thirteen major projects on the Narmada. Only one was in Gujarat: Navagam. The others were all in Madhya Pradesh. It also made two other suggestions that have echoed through the years. First, that the height of the Navagam dam should be 500 feet (and not 320). Second, that irrigation should receive priority over power.

'The water going waste to the sea', Khosla also wrote, 'should be kept to the unavoidable minimum.'

While Gujarat seemed to have liked this particular proposal—meaning no disrespect, authorities in Gujarat have tended to like most proposals to dam the Narmada—the other two states rejected it.

In 1969, the Central Government set up the Narmada Water Disputes Tribunal (NWDT) to find a final solution to the tri-state wrangling over damming the river. The NWDT took a whole decade over its deliberations. One aspect of its work, in particular, has had far-reaching effects. This is its determination of the quantity of water available in the river at the Sardar Sarovar dam site. As you can imagine, this is a vital piece of information. It has bearing on the height to which the dam is built, the way water from the reservoir will be distributed once it leaves the dam, how much water will be available across the command areas, and much more. So it was important to get an accurate estimate.

To start with, the Government of India asked the Tribunal to determine how much water would be available at the dam site at 75 per cent reliability. That is, given the seasonal and yearly variation in water flow, what is the amount of water that can reliably be expected in the river in three of every four (thus 75 per cent) years? Estimating this figure was a major task set before the Tribunal.

Not that it was the first body to do so. That distinction went to the Khosla Committee, in 1964. When it was studying the Narmada, there was actual data available— that is, observed and measured figures—on water flow only for the years 1948 to 1962. Aware that a mere 15 years of data would not be enough to produce a reasonable flow figure, the Khosla Committee used rainfall data dating back to 1891 to estimate water flow. That is, they assumed a relationship between rainfall and water flow and *hindcasted* estimates for the years between 1891 and 1948. Putting the estimates and the actual data together, the Khosla Committee came up with a 75 per cent reliability flow estimate at the dam site of 28.92 million acre feet (MAF), or about 35.6 billion litres. In other words, they estimated that in three of every four years, 28.92 MAF of water would flow past the dam site.

Gujarat accepted this figure. But there was a surprise in the works. Using the very same figures (the 1891-1948 estimates and the 1948-1962 data), Maharashtra's officials produced a figure of 27.17 MAF. Madhya Pradesh, 27.14.

These discrepancies, minor though you might think they are, were one reason the Khosla Committee's proposals for the Narmada died a quiet death.

When the Tribunal sat down to repeat the calculations, it had another eight years of measured data (1963-70) to work with. Using only the twenty-three years' worth of actual figures, without any hindcasting, the Tribunal estimated the 75 per cent dependability flow at 22.6 MAF—significantly below the earlier estimates. The 'benefit' of hindcasting became obvious when a group of experts that the Tribunal appointed used the 1948-70 figures to hindcast back to 1891 again, and then produced a figure of 27.22 MAF.

As Vijay Paranjpye described all this confusion:

> The [phenomenon of varying figures] illustrates three points:
>
> 1. Even using essentially the same series [of data], different authorities have come up with significantly different yield estimations.
>
> 2. As data for additional hydrological years become available, the estimates of yields reduced substantially.
>
> 3. The actual data give significantly lower estimations compared to the hindcast series. (Paranjpye: A-54)

Still, all these efforts counted for nothing. Paranjpye tells us what happened on 12 July 1974:

> The states *agreed amongst themselves* that the yield available for use at Navagam [at 75 per cent reliability] should be taken as 28 million acre-feet [MAF]. The states also agreed that Maharashtra and Rajasthan be allotted 0.25 MAF and 0.5 MAF respectively for use in their territories. [The states further agreed] that the Tribunal should only decide:

a) The allocation of 27.25 MAF among Gujarat and Madhya Pradesh.

b) The height of Navagam dam, and

c) The level of the canal taking off from Navagam dam. [Emphasis added.] (Paranjpye: A-53)

By just such caprice do some things come to be set in stone. It is almost amusing to see the earnestness with which the 28 MAF figure is quoted today, particularly by the builders of the dam. As if it had been diligently calculated, with careful scientific instruments and precision. No, it was really just imposed on the NWDT, and thus on the up-and-coming project.

Another intriguing battle fought before the NWDT was for the height of the dam. From the beginning, Gujarat's whole intent was to get the NWDT to sanction as high a dam as possible. Their original proposal was for a height of 530 feet (Babubhai Jashbai Patel: 13). Paranjpye suggests that they made a 'shrewd calculation that even after the Tribunal reduces the height substantially (say by 80 feet or so), a very high dam could still be built.' (Paranjpye: A-70)

As we have seen, the Khosla Committee, mindful of the problems of R&R and thus the need to lessen submergence, recommended a reduced height of 500 feet. This was the generally accepted figure at the point the NWDT began its deliberations.

A senior irrigation consultant, who appeared before the Tribunal on behalf of the state of Madhya Pradesh, told me that Gujarat was insistent on the 500 feet figure and very unwilling to consider any further reduction. Their claim was that if water had to reach the furthest reaches of the state, the dam had to be this high. (Contrast this with an interesting nugget I came across: a panel of consultants submitted a report to the Ministry of Power and Irrigation in April 1960 that said water could reach Kutch and Saurashtra with a dam just 320 feet tall, the height in the first CWPC proposal. [Paranjpye: A-19]) The state was so firm about the 500 feet height, this official told me, that he, and the officials from other states who appeared before the NWDT, believed the 0.50 MAF allotted to Rajasthan was just 'a ruse by Gujarat to make its case for taking the canal to northern Gujarat. After all, what would Rajasthan do with so little water anyway?'

Still, after hearing the arguments, the Tribunal decided on a height of 455 feet. That occasioned this comment from the former chief minister of Gujarat, Babubhai Jashbhai Patel:

> [T]he Tribunal further reduced the height of the [Sardar Sarovar] Dam to 455 [feet]. . . . This award further reduced the submergence of forest, cultivated and other land of Madhya Pradesh, Gujarat and Maharashtra . . . with a view to reduce the environmental damage to the minimum possible. This obviously will leave room for floods and wastage of precious waters, but would reduce the hardships to the Project Affected Persons (PAP) to the minimum. (Babubhai Jashbai Patel: 13-14)

Note the mention of 'wastage of precious waters', a theme that, as you will see, runs through the arguments for the dam. Note also the undercurrent of disappointment in these lines. Sure, 'hardships' among PAPs would be reduced 'to the minimum', but that would still 'leave room for floods'.

In any case, the NWDT Award of 1979 became the basis of the entire plan to use the waters of the Narmada before they empty into the Gulf of Khambat. The plan is that over half a century, a host of dams will spring up on the Narmada and its tributaries: 30 major ones, 125 medium, and over 3,000 minor projects. (Paranjpye: 1)

With just this plan, the NWDT Award would be an achievement. But it is vital to recognize it is more than that. The Award details not just issues of design such as the flow of water and the height of the dams, not just the ways the states will share benefits from the dam such as electricity and water for irrigation. No, it also spells out, as the Morse Report, which you will read more about later, told us,

> the benefits and procedures for resettlement of those persons in Madhya Pradesh and Maharashtra to be displaced by submergence in the reservoir area, and apportioned the cost of their resettlement to the Government of Gujarat. (Morse: 5)

(Though, oddly, 'the Tribunal did not deal with resettlement of people in Gujarat to be affected by the Projects'. [Morse: 5])

I want to stress the NWDT's examination of R&R here for a very specific reason. Much has been made in the two decades since the NWDT Award about its sanctity, about how various decisions it made cannot be challenged (though as you will see, the Award explicitly does allow for changes). Let's assume for the moment that, as dam proponents often say, the Award is indeed set in stone.

Why then the continuing furore throughout the Narmada Valley, as you will read later in this book, about R&R? Why the dissatisfaction on this vast scale? Is it really possible that all those protests and complaints are motivated and therefore meaningless? If not, why is it that the NWDT's instructions about R&R have been ignored, unlike its views on water flow and dam design?

As Morse says on this point:

> The NWDT laid down conditions regarding
> resettlement. .... At the threshold of our task the
> issue is whether India and the states [i.e. Gujarat,
> Maharashtra and Madhya Pradesh] have lived up to
> these conditions and requirements. (Morse: 7)

Indeed. Much has been written about the effects of dams:
economically, ecologically, agriculturally. Many of those
arguments have been made against the dams on the
Narmada, and many people have rebutted them as well. (I
don't intend to, nor even have the knowledge to, discuss
them in detail here.) But in the end, a major test for these
dams is how far the commitments to R&R have been met.
If they have not, or have been met only in part, that itself
must raise questions about this business of building dams.

And the evidence is that the commitments to R&R, as
spelled out in the NWDT Award, have in large measure
been left unfulfilled.

•

Full-scale work on the dam began in about 1987. Twenty-
six years earlier, in 1961, some 2,000 villagers had been
shunted out of Kevadia, adjacent to the dam site, so that
a colony could come up there for the engineers who
would build Sardar Sarovar. Now many more thousands
faced similar displacement. But this time, the dam's
reservoir would submerge their homes and land. Right
from the beginning, these people were told that their
sacrifice would benefit many other people. Here's how
Morse explains this:

> [The] proponents [of the dam] say that it will bring
> enormous benefits to millions at a cost of displacing
> comparatively few people. [They] say that it will

provide drinking water to over 40 million people and irrigation to 1.8 million hectares of land in that state, not to mention hydroelectric power. They ask that these benefits be weighed against the numbers of people who may be adversely affected. They point out that the majority of the persons to be displaced are tribal people whose lands are said to consist of steep, rocky ground and degraded forests. The land which will be lost, they say, is of marginal value. The juxtaposition of these numbers, of beneficiaries on the one hand and of the persons to be displaced on the other, is said to be sufficient justification for the Projects. (Morse: 5-6)

How should we react to such reasoning? This is a question that has bothered me through all the years I have paid attention to the Narmada issue. I'll attempt some answers later in this book. For now, the important thing is that this logic was put forward as a justification for displacement, for building the dam. Again, note that inherent in this act of weighing people against each other is the political nature of the decision to pursue the project.

•

With previous dam projects, and, in fact, most large-scale development projects in India, people to be displaced were simply displaced. As the plans for damming the Narmada took shape, there was no reason to expect anything but the same story to unfold. Sure, the NWDT Award spelled out generous R&R measures, but the record of dam-builders over the years counted for more than a plan on paper.

Medha Patkar, a young doctoral student at Bombay's Tata Institute of Social Sciences (TISS), was doing fieldwork on tribals in north-eastern Gujarat in the early 1980s. She sought to find out how the proposed dams on the Narmada would affect the lives of these people. It quickly became obvious to her that their lives would change catastrophically. As with previous projects, those building the dams did not seem to have the desire to resettle and rehabilitate the thousands of men, women and children who would be uprooted.

Working with these tribals over the next two decades, Patkar helped them articulate their demands and fight for their rights. The Narmada Bachao Andolan ('Save the Narmada Movement') began by asking for adequate and effective R&R. And while in some ways, the Narmada projects have spelled out on paper some of the best such measures in our history, in practice they have been ignored and flouted.

But quite apart from R&R, the NBA soon came to believe that the very basis for the dam projects was flawed; the very model of development they represented, a gigantic mistake. Could hundreds of thousands of people be legitimately asked for enormous sacrifices to further such a model? Was the national interest—never to be questioned, always the proffered reason for this kind of development— really being served?

The years of confrontation between the NBA and the authorities who wanted to get on with building the dam have been an embodiment of just these questions. In many ways, they have also reflected the evolving view of dams around the world. Once seen as the answer to problems of water scarcity, floods, electricity and so on, many now believe that dams cause more problems than they solve.

This is not to say that dams have not brought enormous benefits. In India, for example, we owe the transformation of Punjab by the Green Revolution to the Bhakra-Nangal dam. But that's just the point. When Bhakra-Nangal was built, and for years afterwards, dams were indeed Nehru's 'temples' of the modern world. Decades later, as reservoirs silt up, as irrigated lands become waterlogged, as we have become more aware of the less desirable things dams bring in their wake, and as some even question the virtues of the Green Revolution, Nehru's vision has started looking distinctly ragged.

It wasn't that Nehru was deluded; nor were we when we looked up in awe at these dams. Rather, it is only now that we are beginning to understand all the costs of building such dams. From waterlogging to the evidence that dams can induce earthquakes, from the hardships of R&R to massive cost overruns—these are all now truths about dams that are widely accepted. It has taken a generation to realize that dams are not—technically, economically, socially, even scientifically—the unvarnished triumphs we once thought they were.

Nor is this happening only in India. In every part of the world, people are questioning every aspect of dam-building, coming to understand what it has meant all these years. As Patrick McCully writes:

> [E]ven in supposedly advanced democracies, dam-building agencies have for years insulated themselves from public control and avoided independent scrutiny of the assumptions used to justify their projects. [The] poor economic performance [of large dams] has invariably been obscured behind a veil of public subsidies. ... Since the early 1990s, however, [this veil] has started to be lifted as governments have attempted to attract private investors to pay for their dams. Private investors need to be convinced that large dams are secure and profitable investments— and the dam industry is being forced to reveal that largely they are not. (McCully: 261-2)

And why are they not? Because dam-builders consistently underestimate—purposely or otherwise—costs, especially R&R costs, and inflate—purposely or otherwise—benefits. In 1991, the World Bank's *Irrigation Sector Review* for India pointed out that bureaucrats commonly 'ensure project acceptance by inflating benefits and underestimating costs'. McCully quotes a World Bank sociologist's account of a dam project he was involved with. The agency building the dam pressured the foreign firm preparing a feasibility study to greatly reduce the estimate of people to be displaced or eliminate it completely. It 'was concerned that if the real scale of the displacement were to become known, internal political support and/or external financing of the project would be doubtful.' (McCully: 305)

All of which only underlines the real reason dams are attractive: they are political goldmines.

The NBA may have lost the case in the Supreme Court—you will read about that in a later chapter. But what it has achieved over the years is to bring these issues about Indian dams into the debate—and, in fact, the debate today goes beyond dams to our very model of development. It's not clear that India is turning away from that path of development, but clearly questions about it will now always be raised. People threatened with displacement will always demand their rights, fight for justice. It remains a hard fight, but it's better, at any rate, than being voiceless as in the past. Indeed, it's their best guarantee of a better future.

2

Why Build Dams

By now, you know what I feel about the Sardar Sarovar dam. Yet, as I have indicated earlier, many people in Gujarat and elsewhere believe passionately in that dam. What makes them do so? Why do they think it is a solution, even the only solution? These questions fascinate me, driving me to find answers.

In my experience, when asked to list the benefits of a dam, the first thing people mention is usually electricity. 'What a pity that the Sardar Sarovar dam is so far behind schedule,' a friend at the World Bank once told me, 'because it would have been generating electricity by now. And you know, hydroelectricity is the cleanest form of electricity we have.'

Maybe, but what's interesting about Sardar Sarovar is that it was never meant to be a significant long-term source of *electricity*. (Recall, for example, the Khosla Committee's insistence that irrigation should receive priority over power.) Sardar Sarovar was always intended as a solution to the *water* problems of Kutch and other 'drought-prone' areas of Gujarat and Rajasthan. For years, its builders have called it the 'only solution' to the water problems of these areas, the 'lifeline'—you've heard that term already—of Kutch and Saurashtra and Gujarat itself.

This is worth remembering because the success of the dam must then be measured in terms of how it actually delivers water. Nothing else. It is water that has defined the height of this dam. And water is a particularly important issue in Gujarat.

•

Just as the word 'temple' comes up while discussing dams, so does the idea of 'waste'. As in, how can we waste the water in our rivers by letting it flow into the sea? The world over, as long as dams have been built, this has been a persuasive, eloquent and hard-to-refute argument. 'Pity us,' said Joseph O'Mahoney of Wyoming during a heated 1940s debate in the US Senate on damming the Colorado River (by now perhaps the most dammed strip of water in the world):

> Let us store the rainwater which for thousands of years has been rolling down the Colorado River without use. Please have some pity on the area, which is the arid land area of the country. It wants to conserve the great natural supply of water which the Almighty placed there, for man to use, if he has the intelligence and courage to use it. (Reisner: 149)

The powerful image O'Mahoney was conjuring up—of water flowing past thirsty Americans 'without use'—was, as I said, hard to refute. It is no accident that he invoked 'the Almighty', that he used such inspiring words as 'intelligence' and 'courage' to describe possible efforts to tap this water. Stopping the waste was America's divine duty, a supreme human endeavour that would indeed take intelligence and courage to complete. In any case, it was also politically wise for O'Mahoney to put it that way, because language like that buys votes.

Besides, O'Mahoney was up against Paul Douglas, a wily senator from Illinois. Douglas had just raised a troubling question about dam projects on the Colorado River: why should the nation lay out what amounted to $2,000 an acre to build these dams if the irrigated lands they created would sell for only $150 an acre? Faced with this calculation, O'Mahoney produced his trump card: the need to save the water that was 'rolling down . . . without use'. What were mere dollars in the face of that almost ungodly waste?

The Narmada river has been mute witness to such logic as well. Why should it deliver its bounty fruitlessly into the Gulf of Khambat when the precious water can be used elsewhere? (Recall, once more, the Khosla Committee's insistence that 'the water going waste to the sea . . . should be kept to the unavoidable minimum.') This is an important ingredient in the arguments for building a dam. Here are three excerpts from SSNNL leaflets that make the same point:

> Since time immemorial, over 96 per cent of the Narmada waters go waste to the sea unharnessed occasionally spelling disaster to the life, property, cattle and agriculture produce of the people living by the banks of the river. . . . There is only one way to save Gujarat . . . harness the 45,000 million metre [sic] Narmada river for public use, whose waters go waste in the sea and thus turn the dry land into the proverbial 'NANDAN VAN'. (Jhala)

> [The] Narmada waters, only 4 per cent of the water potential of the entire basin of which is being utilised

at present, are free to flow to the sea [when] millions
of human beings and animals pass through severe
scarcity conditions for years continuously just because
there is a continuous failure of rains in several parts
of Gujarat. (Babubhai Patel: 11)

Till now, the waters of the Narmada have been
discharging themselves without let or hindrance into
the Arabian sea: there could have been no question
of utilising them during the hundreds of years of
foreign rule. [*Why?*]. After the arrival of Independence
in 1947, Government of India became alert to this
problem [and] started investigations. [Interjection
added] (Dalal)

Of course, if there was no need for water in Gujarat, this
argument would—so to speak—hold very little water. That
is, the flow of the Narmada into the sea is seen as a waste
precisely because there are arid areas of the state where
people need water badly. Just as with the Colorado River.
So the case for the dam has always referred to the troubles
caused by aridity. Here are a few lines from another
booklet that has been circulated in Gujarat:

[T]here is urgent necessity for sustained development
of Gujarat and also for improving the living conditions
of people of Gujarat. Increasing agricultural
production and reducing hardships for common
people in regard to drinking water and water for
household use cannot be valued only in terms of
rupees but have to be considered as a measure which
must be taken to alleviate the difficulties faced by
people in the lower economic strata. (Amin and
Naik)

Did you catch the faint echoes of O'Mahoney? Like him,
in his response to Senator Douglas's calculations, these

writers are trying to place the argument for the dam beyond such considerations as 'rupees'. Indeed, they say, the need to bring succour to millions of suffering Indians does outweigh the costs, *whatever those costs may be.*

And, in fact, the whole thrust of much of the SSNNL's literature about the project is the emphasis on drinking water, even more than irrigation. This is how V.B. Patel of the Gujarat Water Supply and Sewerage Board puts it:

> Sardar Sarovar has been talked about as the lifeline of Gujarat. Most of the people feel that it is called lifeline because it would provide plentiful water for irrigation. . . . But very few know all the facts. One of it and perhaps the most important is that it would provide LIFE GIVING SAFE WATER for the thirsty millions of Gujarat; that it would help reduce water related health problems; that it would help increase productivity through improved health; that this would reduce burden on the State in fighting health hazards; that this would increase the wealth of the State due to increased productivity of the people. WATER IS LIFE. Survival of life is not possible without water.
> (V.B. Patel)

In its essence, this is the case for the dams on the Narmada. This double-shot vision: of water-starved millions in Gujarat, of Narmada water flowing wastefully into the sea. Every paper, article, booklet about the projects that I have seen starts from this place. Everything else—generation of power, flood-control, fisheries in the reservoir, employment—is secondary to these considerations. This is true to the extent that several of the publicity booklets issued by the SSNNL refer to the SSP as 'a real drought-proofing project', and in so being the 'lifeline of Gujarat'. Or, sometimes, 'the LIFELINE OF GUJARAT'. Even the

Morse Report reflects this emphasis on water over other benefits:

> The Sardar Sarovar Projects are intended to bring drinking water to Kutch and other drought-prone regions of Gujarat, and to irrigate a vast area of that state as well as two districts of Rajasthan.
> (Morse: p. xii)

> Supporters of the dam say that it will provide drinking water to over 40 million people and irrigation to 1.8 million hectares of land in [Gujarat]. . . . The Chief Minister of Gujarat, Chiman Patel, predicts that if the Sardar Sarovar Projects do not go forward, if water for drinking and for irrigation is not made available to the drought-prone areas in Kutch, Saurashtra and northern Gujarat, the result will be [that] hundreds of thousands of Gujarati citizens . . . will be forced by drought to migrate from their homes. (Morse: 5-6)

Note that Chiman Patel does not speak of the consequences of *power* not being made available, only water. Nor does he speak of the thousands of Indian citizens who will be forced by the dam to migrate from *their* homes.

Yet Sardar Sarovar does have other claimed benefits. Actually, the dam will generate electricity, 1,450 megawatts (MW) of it. Project authorities say this will stimulate local industry. They also believe the dam will employ several hundred thousand workers daily while it is being built, and a slightly smaller number after it is built.

All of this, while commendable, really pales into insignificance when compared to the plans for water. In the end, it is water, and how it is delivered where it is required, that defines the project. At the risk of belabouring the point, water must then become the criterion by which we judge it, evaluate the way the dam is being built.

And the literature from the builders of the Sardar Sarovar dam offers many figures to show how the water will be used. According to the SSNNL publication *FACTS: Sardar Sarovar Project,*

> the project with 75 per cent of its command as drought-prone spread over 12 districts, 62 *talukas* and 3393 villages would extend irrigation benefits to 25 lakh agriculturists and hence be a boon to the state.
> ... The problem of domestic water supply to arid areas of Saurashtra and Kutch would be solved permanently. (*FACTS* 1989: 48)

And of course, this is just why the dam is such an emotion-charged issue. People do suffer from the lack of water, especially in Kutch, Saurashtra and northern Gujarat. They want reliable sources of water. Over the course of perhaps a generation, they have come to see the Sardar Sarovar as one such source, even the only such source.

That hope is really the case for the dam.

More learned books will spell out the case for the dam in great detail. Here, on the other hand, I'd like to do it another way: by giving you a sense of how much water is an issue in the dry parts of Gujarat, of how people there yearn for it every day, of what a preoccupation it is for them.

•

I got a taste of the depth of people's feelings about water one sunny afternoon in Toraniya. Days after the enormous Kutch earthquake of 26 January 2001, I went to this tiny, devastated village as part of a team that stayed there a week. We had relief material and did what we could to help people rebuild their lives. But we were also trying to understand the villagers' concerns, beyond the immediate task of coping with the havoc the quake caused.

Seeking to learn more, a few of us filled our water bottles, stuck hats on our heads, and joined Lirabhai on a brisk walk through the fields east of Toraniya.

Lirabhai grew up in a farming family in Toraniya. Now, he spends some months every year working in a STD booth in Bombay. He needs to do this because there isn't enough water available in Toraniya for him to farm through the year. But when he is there, Lirabhai spends much of the time trying to get fellow villagers interested in his pet idea.

Lirabhai wants to create a *talao*, a pond or reservoir, that will be a source of water to the village all through the year.

Our brisk walk takes us about a mile east of the village, to the bed of a stream. It is dry now, but during the monsoon it is full. Lirabhai's dream is to build a small dam across the stream to make his talao. 'It will submerge some public land, some belonging to five other families in the village, besides some of mine,' he tells us. 'But it will help the village, and it will help me, so I don't mind giving up my land. The other families have also agreed to contribute theirs.' He is now working on how to get permission to submerge that 'public' land.

So we stand on this gentle rise above the bone-dry bed of the stream, wiping the sweat from our brows, trying to visualize a large body of water where we see only scrubby bushes, sandy soil and rocks. It's hard, but it's just as hard to ignore Lirabhai's enthusiasm and optimism.

It strikes me at that moment: quake or no quake, once the immediate task of rescuing people and treating injuries is over, people in Kutch will return to what must be an old preoccupation here. Water.

Take the sheer irony of it. In this Kutch village are a few hundred of the very people who have been persuaded over a generation that a huge dam on the Narmada will bring them water. People in whose name it is being built. And come to camp in this Kutch village, to help quake victims, is a team made up largely of farmers from the very area—Nimad in Madhya Pradesh—that will be seriously affected by that same dam on the Narmada.

Solely because they wanted to do their bit to help the families hit by the quake, these Nimad farmers collected several dozen large sacks of grain and a few hundred plastic tarpaulins, gathered many unwieldy bundles of clothes and quilts, steel utensils, candles and the like; piled everything into three hired trucks and a jeep; secured it all with strong ropes, climbed on top themselves and drove for over forty-eight hours—two fiercely sunny days and three chilly nights—to reach here. I joined them soon after they crossed into Gujarat.

Really, think for just a moment about what happened in quake-wracked Toraniya: farmers facing the prospect of homes lost to a dam voluntarily bring succour to other farmers who are told that that same dam will one day deliver precious water to them.

The villagers here seem grateful that we have not just driven in, flung some clothes or biscuits from atop our trucks, and driven on. We spend several days here, build friendships with several villagers, Lirabhai included. After taking time to understand the village dynamics, we buckle down to work where it is most needed. The school is rubble; so with material left behind by a team from Delhi, the Nimad farmers erect a large tent in its compound. The schoolchildren gather again that very evening; classes resume the next morning. Outside, the panchayat office is also shattered. Another tent goes up there, and village meetings resume too. In groups of two or three, we fan out across the village, helping shelterless families put up tents with the large plastic tarpaulins they have received. I tag along with Jitendra, the strong young son of a farmer from Nimad. His efficiency and good cheer, as he swiftly erects a tent for Chhaganbhai's family, make me feel clumsy, even redundant.

I mention these details because through it all, water remains a major preoccupation. Most houses here have pipes and taps that used to bring them water from a bore well near where we are camped. But with no power supply since the quake, the pump cannot be operated. There's no way of telling, until the electricity returns, if the well is damaged. But even if it isn't, what about the pipes and taps? Did they survive the quake? In several houses, the pipes have been twisted and even broken.

In this situation, people here have two main sources of water—water tankers that come once a day, and a well outside the village from where water must be drawn by hand. So, with a couple of villagers, some of our team members go into Bhachau—headquarters of the earthquake relief operations—to arrange with the authorities for a large plastic tank to store water in the village itself. It arrives in Toraniya at two in the morning. Much old-fashioned Nimadi and Kutchi ingenuity is needed to fit different bits of pipe together so the tank can be filled from the bore well.

But I am there when the electricity returns and the pump is turned on, when water begins to flow into the tank, when the line of taps it feeds is turned on, when people race from all over the village, whooping in delight, to fill their pots. The relief and joy are palpable, and we outsiders cannot resist cheering either.

We now understand Lirabhai's dream. We understand what a large talao like that, with the water it would hold year-round, would mean to this village.

It makes me think: in flesh and blood, embodied in the slender figure of Lirabhai, hinted at in his dream, this is the case for the Sardar Sarovar dam.

Yet that hasn't stopped him from welcoming these farmers from Nimad to his village, from explaining his thoughts to us. He believes implicitly in Sardar Sarovar, but he knows it will deliver water here only years from today. In the meantime, he says, he and his mates must find solutions to their water needs *now*. Today. Once they are over the immediate shock of the quake, that urgency has returned, sharpened by the possibility of damage to their well.

And that's why Lirabhai takes us to see the lay of the land he has in mind, talks about his idea of a talao. That's why he appeals to these farmers from the 'other side'—in every sense of the phrase—of the Sardar Sarovar dam. He knows that water concerns them just as much it does him. Their visit and stay here is proof of that.

Only, it had to take an earthquake to bring these people from these two 'sides' together.

•

My week in Toraniya taught me a simple lesson. It's easy to find and read any number of books or papers or SSNNL brochures that spell out why a dam—this dam—should be built. But nothing beats a visit to the people who long for that dam—or, more correctly, for the water that dam promises them. The plight of quake-hit Toraniya, the way water dominated thought and action, was in some ways the clearest statement I had ever come across of the need for the Sardar Sarovar dam. That is, a need for the dam as a source of water *now*.

Toraniya also showed me the futility, the tragedy, of the polarization around this dam. Sitting and chatting together in front of me were small farmers from both sides of the debate, sharing as they never had done before—as they had never had the chance before—their understanding of water. I think the Nimad farmers went home with an appreciation for Lirabhai's difficulties, for his efforts to build a permanent store of water in Toraniya. I think Lirabhai and his villagemates understood that these Nimad farmers were not some eco-terrorists who were bent on depriving them of water.

If only there could have been more such meetings across the divide of the dam. Rather than the black-and-white caricatures we have today, we might have understanding. And that understanding might be a fuller, deeper, more complete case for a dam. That is, and I cannot emphasize this enough, a case for a dam seen as a source of water *now* to citizens who have scrabbled for it for years.

And it is in that light that we must evaluate the dam on the Narmada that promises so much to Gujarat.

3

From the Horse's Mouth

Yes, I do believe the Sardar Sarovar dam is a gigantic mistake. On balance, I have come to believe dams, and this one in particular, have major problems. I say 'on balance' because I am willing to recognize that there are others for whom the balance tips the other way. There are dams that have been built conscientiously, without flouting requirements, and have delivered on their promise. While it is important to understand the inherent flaws in dams, it is just as important to recognize that some have brought benefits.

The educationist Krishna Kumar once wrote in the *Times of India*:

> [T]he whole Narmada issue should form the cutting edge of any knowledge in relation to learning at any school or college in the country. ... Economy and planning, ecology, ethics and the state are all intertwined in the drama unfolding in the Narmada Valley. And it is no ordinary sociodrama; it has an epic scale and pervasive implications for our democracy. (Kumar)

I believe he was referring precisely to the need to understand all that dams entail. If the Narmada controversy is to have meaning for, and lessons for, our democracy, we must know all we can about it. In that spirit, this chapter will examine some aspects—some eye-opening aspects, at least to me—of the planning and construction of the Sardar Sarovar dam.

It's true, my belief about the Sardar Sarovar dam is that the balance is, and has always been, tipped one way. Not out of any intrinsic faults or benefits with dams, but because of the way this dam has been planned and constructed.

That *process*, I believe, has what Krishna Kumar calls 'pervasive implications for our democracy'.

•

After the NWDT Award of 1979, the governments concerned applied to the World Bank to help fund Sardar Sarovar. The money became available in 1985, when the Bank gave India and the governments of the three states $450 million in aid to be used for the project. (This amount 'contributed 18 per cent of the cost of the dam and power project and about 30 per cent of the water delivery project.' [Morse: 5])

Both the Award and the Bank agreements spell out environmental and R&R requirements that the project would have to meet. In fact, the Bank's policies were based on the Award's, though the Award's views on R&R were restricted. As Morse points out,

> The Tribunal had been convened to adjudicate an interstate dispute and apportion benefits and costs. . . . It was not intended, however, and should not have been expected, to establish policies and programmes that would meet the needs of the affected

people of the whole complex of Sardar Sarovar
Projects. It did not even mention the Gujarat oustees,
nor did it concern itself with the people potentially
affected by the canal and associated irrigation works.
It did not take into account the cultural attributes of
the oustee population; there is no discussion of tribal
peoples, encroachers, or the meaning of 'landlessness'.
(Morse: 48)

Morse observes that these characteristics of the Award
make it unsuitable as a basis for R&R. What's more, this
comment hints at the continuing confusion about who is
actually affected by the dam. This is no idle oversight. The
stakes are high. Numbers matter. The definition of exactly
who is affected by the project can make the difference
between a manageable number of people who need R&R,
and a quite unmanageable number.

And that raises questions such as: Is it only those in
the submergence areas who are to be treated as Project
Affected Persons? What about those who give up their
lands so the canals can be built? Should the families who
were displaced to build the dam authorities' colony at
Kevadia be considered as project-affected?

As you will see, the Supreme Court has refused to
recognize people displaced by the canals as project-affected.
There have been wrangles about the status of people
described as 'encroachers' on land to be submerged, of so-
called 'major sons', of those Kevadia oustees, and much
more. So if you believe the term 'project affected' is
simple enough in its meaning—that is, anyone affected by
any aspect of the project—think again.

But whatever that definition may be, much has always
been made of what each PAP will get as their R&R
package; we are also reminded that the terms of this R&R
package are the most generous ever offered in India.

Before construction began, Gujarat had 'liberalized' its R&R package beyond the parameters spelled out in the Award. Here are some of its important ingredients:

1. Agricultural land equal in area to the land acquired, subject to a minimum of 2 hectares (ha) per project affected family.

2. Landless agricultural labourers in the submergence area would also be entitled to 2 ha.

3. Encroachers on both Government and forest lands acquired for submergence would also be entitled to 2 ha.

4. Sons over the age of 18 ('major' sons) are to be treated as separate families and are thus entitled to 2 ha of agricultural land.

5. Every family will get a 500 square metre residential plot free of cost.

6. Every family will get a grant of Rs 10,000 to construct a plinth on their residential plot.

7. A subsistence allowance of Rs 4500 per family.

8. A grant for ploughing fields of Rs 600 per family.

9. A grant for buying livestock and agricultural implements of Rs 7000 per family.

10. Roof tiles equivalent to the roof area of the house, subject to a maximum of 85 sq m.

11. Free transport of dismantled building materials from the old house to the new site.

12. Insurance covering the homes, contents, injury and death.

In addition, the package promises other benefits to the resettled community: an approach road, internal roads, a building for a primary school, health centre, children's park, wells, electricity, a village pond, dispensaries and more. (*FACTS*, 1998)

No doubt the package, as spelled out here, is indeed a thoughtful, comprehensive one, down to the roof tiles and insurance coverage. Project officials always point to this package (and similarly liberal ones from the Madhya Pradesh and Maharashtra governments) as proof of their concern for PAPs.

Still, recall the observation that Sanat Mehta, once Chairman of the SSNNL, makes in one of his Nigam's booklets: '[T]his is the first ever project of the country wherein rehabilitation problem was considered in much details [sic]. . . . [If] the project would have been executed in the old fashion, its cost would have been far less.'

This is merely official confirmation of what critics have said for years about development projects in India: that they have never bothered about the people they displace. (If they *had* bothered, the 'cost would have been far' more than it was.) In the SSNNL booklet about R&R, this is repeated: 'So far rehabilitation of affected people of irrigation and hydro power project [sic] was considered incidental to any development Project and not an important, separate activity.' (*Rehabilitation & Resettlement:* 5)

It is precisely this failure to carry out R&R in our development projects through the years that makes many people immediately suspicious of how it will happen on the Narmada. No matter how thoughtful the new packages are. For the people who will be affected, past experience carries substantially more weight than the mere existence of new measures.

What must concern us, then, is how the packages are

implemented. That is, how many people actually go through the process of R&R and enjoy the benefits? How satisfied are they with their changed situation? After all, as the same *FACTS* booklet tells us, 'the primary objective of the policy is that the economic condition of the PAPs must improve significantly after resettlement.' (*FACTS* 1998: 34)

As you will see in different places in this book, project officials cannot honestly claim that such improvement has happened. I myself have met enough people whose experience of R&R has been bitter.

But if you are willing, as a lot of people are, to discount such people as being congenitally dissatisfied, or even instigated to complain, you should take a look at the project authorities' own plans for R&R. They offer clues to understanding what has happened with R&R. And why.

Note: You will have noticed many references to SSNNL booklets published in about 1988-1989. Why those particular booklets, you might ask, aren't they outdated? I quote them particularly because they reflect conditions and plans at about the time construction on the Sardar Sarovar dam started. They raise the questions: What had the authorities planned at the start of the project? How did they visualize the progress on it?

Seeking answers to these questions, I read the SSNNL booklets carefully. Only to find myself caught instead in a jumble of figures. (See Appendix A for more).

•

Let's start with the number of people the Government of Gujarat (GoG) estimates will be displaced by the Sardar Sarovar dam. In the 1989 edition of the *FACTS* booklet, I read that the dam 'will displace about 100,000 persons of 248 villages of the States of Gujarat, Maharashtra and MP.'

In the 1998 edition of the same booklet, the various figures are gathered together in a convenient table. Poring over it, I find that the number of villages has been revised slightly downwards: it is now 245. But the number of PAPs? 40,727.

What happened to that figure of 100,000? Put it another way, where did 60,000 people vanish to?

How is it possible to come across this discrepancy and not think, what kind of idea do the authorities themselves have of the numbers involved? After all, edition after edition of this booklet has been authored by the same officials all these years, and published by the same organization. Do they know of this discrepancy, and if so, why it exists? One explanation, perhaps, is that in 1988, unlike a decade later, accurate surveys had not yet been done. But then how did they arrive at the figure of 100,000 at all? How is it possible for even an estimate to be so far off the mark?

With your mind buzzing over these puzzles, you dig some more. The figures make you wonder if they have simply been taken off the top of somebody's head. Take what else I found, all still from SSNNL booklets. An October 1988 booklet written by Sanat Mehta tells me that the Sardar Sarovar reservoir 'will displace 67,000 persons of 237 villages of the states of Gujarat, Maharashtra and MP.' Lalit Dalal's SSNNL booklet, from the same time, says that 'the SSP will displace about 88,000 persons of 248 villages.' The one by B.K. Jhala quotes in full an October 1988 editorial in the *Business Standard* that says 'estimates are that some 170,000 people will suffer because of the submergence of their lands and dwellings by the reservoirs.' A SSNNL leaflet I have repeats that 237 villages will be affected, and says the 'number of families affected' will be 10,750. Multiply that by five, the generally accepted family size in such calculations, and you have 53,750.

Now I have not just two, but *six* different numbers. How did each author arrive at his particular figure, I wonder. The SSNNL published all these authors' efforts: did they not even stop to compare notes? Did the SSNNL not have some 'official' estimate of its own for displaced people that could have been furnished to these writers? And if not, what kind of conclusion should we reach about this project?

With all these numbers, it is almost a relief to open the undated, but likely also 1988/1989, booklet from the SSNNL titled *Rehabilitation and Resettlement*. Because there I find this figure: 66,675. At least it essentially agrees with one of those other numbers (Sanat Mehta's 67,000).

Now consider the figures the same SSNNL booklets list for the other side of the dam: the area it is supposed to irrigate. Whether for 1988, 1998 or in between, the numbers are remarkably consistent: 1.793 million ha, 1.792 million ha, 1.7 million ha, 1.8 million ha. Though oddly enough, it is the *Rehabilitation and Resettlement* booklet that has a wildly different figure: '3.45 million ha . . . will get irrigation facilities.' Perhaps this should be treated as an aberration, a misprint.

Why do you think the authorities' figures for irrigation are in such close agreement, but the figures for the numbers to be displaced vary so dramatically? The only explanation I can come up with is what I have suspected all along about this project: the benefits are given a great deal of attention, but the costs are treated as a minor irritant to be ignored as far as possible. In particular, R&R is not to be taken seriously, certainly not seriously enough to estimate affected people properly. Not only do the booklets reveal a reluctance to estimate the numbers of displaced people accurately, they also show a supreme confidence that nobody will actually peruse them and find these varying numbers.

•

And it goes well beyond R&R too: more than anything else, the SSNNL booklets are a minefield of oddities like this. They confirm another of my pet theories: that it is not the critics who make the best case against Sardar Sarovar, but the men who are building it, who write booklets to defend it.

Consider two more interesting features, extracted at random, from these defences.

Now I am particularly interested in learning about the benefits of this project. Which is why, in every bit of literature I find about it, I turn first to the description of benefits. This is what I looked for in the SSNNL's 1988 booklet authored by its then chairman Sanat Mehta. Mehta writes that if the dam is built, '*2.01 crore [20.1 million] population in 2021* will be able to get drinking water.' [Emphasis added] (Mehta 1988: 5)

This claim, by itself, makes me slightly uneasy. When this book was published, 2021 was 33 years in the future. Claims for a generation ahead, as I've noted before, seem to me to be effectively meaningless. Yet they are routinely held up by project authorities as the reason for the dam.

But is it legitimate at all to use a 2021 figure here? I may have thought so, actually, until two things happened.

First, in the 1989 *FACTS*, under the heading 'Benefits', I ran across this sentence: '[The dam] would thus cater for the domestic water supply needs of about *295 lakh [29.5 million] persons (2021 Census).*' [Emphasis added]

Odd, I thought. Between 1988 and 1989, the estimate for the population in 2021 rose from 20.1 million to 29.5 million. How did that happen? And this 29.5 million figure is attributed to an event—the 2021 Census—*that hasn't even happened yet.*

Second, in the 1998 edition of *FACTS*, also under the heading 'Benefits' (page 26), there's this: '[An allocation] has been made to provide drinking water ... in Gujarat for present population of 18 million and *prospective population of over 40 million by the year 2021.*' [Emphasis added]

I would love to hear how, in a space of a decade, the estimated 2021 population inflated itself from 20.1 million to 29.5 million (and that in one year) and finally to 40 million.

And there's still more. On page 68 of that very same *FACTS* of 1998 that mentions the 40 million figure, there's this: 'The project will provide water to a projected population of 24.3 million in the year 2011 and 29.26 million in 2021.'

So the SSNNL itself has given me four different figures for the numbers the dam will benefit in 2021: in chronological order, 20.1 million, 29.5 million, 40 million and 29.26 million—and the last two from the same publication.

What happened here? Do the people writing these booklets, or updating successive editions, consider each figure and randomly decide to change them, preferably upwards? Is this more top-of-the-head stuff? Are we, who want to know about this dam, not even entitled to reasonable and accurate estimates of the numbers of people who are to benefit from it?

Then the booklets refer to three years of severe drought in Gujarat in the mid-80s (1985-88, actually). This mention of drought is meant to support the case for the dam. The situation in those years was so difficult, the booklets say, and the GoG had to spend so much and make so many arrangements to provide relief to the victims of drought, that it became clear that this dam on the Narmada was needed, that it would put an end to all the misery.

A good argument, to be sure. And what was the amount the GoG spent during the three years? Sanat Mehta writes that 'the State Government had to spend Rs 200 crore [2,000 million] to ease the serious crisis of drinking water created in last three years by consecutive droughts.' (Mehta 1988: 2). In another booklet (*Sardar Sarovar: Gujarat's Hope*) from the Nigam, I find that 'a staggering amount of Rs 10,000 million has been spent on relief works between 1985 and 1988.' As if that wasn't enough, the 1998 *FACTS* tells me: '[T]here were 3 consecutive drought years from 1985-86 to 1987-88 during which Gujarat spent as much as Rs 15 billion [15,000 million] on drought relief measures.'

Well, which is it? Rs 2,000 million, Rs 10,000 million or Rs 15,000 million? Really, could it have been that difficult to come up with one figure for the expenditure to fight drought in these three years? Or are these more top-of-the-head numbers?

Really, why is there so much evidence that so many aspects of this project have been addressed in a slipshod, half-hearted and incomplete fashion? And given this, how seriously can we take anything about the dam, in particular its ability to deliver what it promises?

•

Then there are claims that are, to put it kindly, misleading. In the 1998 *FACTS* booklet, printed in bold and italics so you cannot miss it, there's this line: '*According to the Award of NWDT, the parameters of Sardar Sarovar Dam will neither be reviewed nor changed till 2025 A.D., i.e. 45 years after the notification of the Award.*' This forty-five years requirement has by itself gained a certain degree of fame, because it is often cited by the GoG to dismiss any call for a review of the project.

This might be acceptable if it were true. In the NWDT's 'Final Order and Decision', Clauses III and IV spell out the allocation of water in the river to the four states; that is, the way that 28 MAF and any excess flow over that amount is to be divided. That is all they are concerned with. Then Clause V, in full, reads thus:

> Our Orders with regard to the equitable allocation in Clauses III and IV are made subject to review at any time after a period of 45 years from the date of publication of the Decision of the Tribunal in the Official Gazette.

This is the NWDT Award's mention of the forty-five year stipulation. That is, it is the allocation of water, and nothing else about the project, that is subject to this embargo. So to claim that 'the parameters of Sardar Sarovar Dam' cannot be changed for forty-five years because the NWDT Award says so is to claim far more than the NWDT ever dreamed of.

And, in fact, despite this Clause V, the Award explicitly allows changes of any sort. Towards the end of the 'Final Order', Sub-Clause 17 of Clause XIV reads:

> Nothing contained in this Order shall prevent the alteration, amendment or modification of all or any of the foregoing clauses by agreement between all the States concerned.

That is, if the states choose to agree, they can change anything in the order. Including the mention of forty-five years.

As we shall soon see, Madhya Pradesh has proposed to make use of precisely these words.

•

The SSNNL booklets are not the only GoG literature that say revealing things about Sardar Sarovar. Let me share here some extracts from a 2001 report on the environment in Gujarat, published by the Gujarat Ecology Commission (GEC) in Baroda. It is one result of the State Environment Action Plan (SEAP), which the report's foreword describes as being 'entrusted to the Gujarat Ecology Commission [by the Government of Gujarat]'. Indeed, the then chief minister of Gujarat, Keshubhai Patel, formally launched SEAP on 15 August 1999 (the foreword carries a picture of him at this launch). And a 'high-level committee headed by the Chief Secretary and representing major Government departments provides policy guidelines.' All this, to emphasize that this effort, and thus this report, has approval and guidance from the highest levels of Gujarat's government.

One reason for preparing this SEAP is clear from this paragraph in the Foreword:

> The true cost of the development in terms of degradation of productive resources, damage to the health of the people and pressures on the life supporting systems has not been precisely documented. It is perceived to be high enough to earn for the state the sobriquet: 'most polluted state.'

The report has this comment on 'Waterlogging':

> Excessive irrigation in command areas of irrigation projects of Ukai and Kakrapar has led to waterlogging and salinity over 0.185 mha. This affects 52.18 per cent of the total command area of 356,080 ha. In the

Mahi command area of 212,694 ha, as much as 138,676 ha representing 65.19 per cent of the command area is salinity affected. Other factors like poor natural drainage and lack of incentives . . . have added to the problem. (SEAP: 23)

You will read briefly about waterlogging and salinity, and the close relation between them, later in this book. What this report is saying is frightening indeed: that vast areas of irrigated land are becoming useless for agriculture, as a direct result of that irrigation. But if it's possible, there's an even more frightening aspect to all this: the blithe treatment of these problems in the *FACTS* booklets. From the 1998 edition, some relevant statements. First:

Effects like waterloging [sic], salinity etc., are unjustifiably ascribed to large dams. These are caused due to poor land and water management which has nothing to do with the size of the project. In India less than 10 per cent of irrigated land is waterlogged. (*FACTS* 1998: 53)

Less than 10 per cent? How then do we account for the 52.18 per cent and 65.19 per cent figures the GEC report mentions?

More alarming, the *FACTS* booklet says:

Water logging even in Ukai-Kakrapar and Mahi Kadana is hardly 2.5 to 3 per cent. (*FACTS* 1998: 56)

2.5 to 3 per cent? Again, how do we account for the 52.18 per cent and 65.19 per cent figures the GEC report mentions? Of all these figures coming to us from the GoG itself, which ones are we to believe?

As if these discrepancies in figures were not enough, consider this confession in the GEC report:

The situation [of scarcity of water and consequent resentment in rural areas] would only *marginally improve* even after ... construction of the Sardar Sarovar project. The project would definitely go a long way in meeting the domestic and industrial requirements of water in Saurashtra, Kutch and North Gujarat. But a glance at the map of the command area of the project clearly indicates that large areas of water deficit regions would not get any water from the Narmada project. As many as 35 out of the 53 talukas having dark or over-exploited status of ground water development are outside the command of the Narmada project. As a matter of fact, Narmada waters will serve only about 22 per cent of the cultivable land in these 53 talukas where ground water is exploited. [Emphasis added] (SEAP: 33)

All of this is what opponents of the dam have been pointing out for years. The NBA's contention has always been that, despite the project's claims that it will take water to needy regions, the project's own plans show that major parts of such regions will not get water. A generation and more after officials in Gujarat began dreaming of this dam, we find admissions buried in this report by other Gujarat officials that only bolster that contention. That admit that the SSP will only *marginally improve* the water situation where water is most needed.

Makes you wonder. Why is this dam being built at all?

Oddly enough, as I have implied elsewhere, that's a question that won't go away even if you are willing to ignore such reports as the GEC's. If the dam authorities, if government officials themselves, are so unclear on every aspect of this project, unclear enough that they give us figures about these aspects which seem to be chosen at random rather than carefully produced, why is this dam being built? What does this monumental lack of clarity say about the way it is being built?

•

Seeking some answers to those questions, and keen to learn about various related issues, I made a trip to meet K.R. Datye. An engineer by profession, Datye spent years intimately involved with the planning and execution of various dam projects, including those on the Chambal, the Krishna, the Tapi and more. He also worked in the Central Water and Power Commission (CWPC). More recently, he has taken substantial interest in irrigation and water policy, urging more sustainable and equitable approaches to these matters. In his capacity as a consulting engineer, interested in water issues, he is an expert member on the Narmada Control Authority's R&R Sub-Group.

Datye is certainly no blanket anti-dam man. 'What would have happened to us,' he asked me, 'if we did not have massive irrigation available in our early years? The Republic would not have survived; it would have been a situation worse than the Bengal famine. Water storage is an essential function for any society, and water is fundamental to raising people's standard of living.' Water from large dam reservoirs, he continued, can be an instrument for social transformation—which is what the Bhakra-Nangal dam accomplished in Punjab. Therefore, the view that all large dams are mistakes is, said Datye, and he repeated this, 'an unreasonable one'. Some days later, he wrote to me: 'My unhappiness with the opponents of large dams is that they are limiting their efforts to a single point agenda of large versus small. Social mobilisation efforts should be extended to land, water and agriculture sector reform.'

Datye raises an issue here that is of great importance.

While we pursue irrigation projects with conviction, we have not pursued land reform with any seriousness: that is, we have not given land rights to small farmers on a large scale. The two cannot but go together. Our political leadership has never cared about this process. Development can only build on such land reform, can only take place when people have a stake in the land. This is the view, Datye told me repeatedly, we must take of R&R. When people are displaced by a dam, they must get clear title to a new plot of land, and access to the water resources that will make it productive. (Giving them cash compensation, he emphasized, is no substitute for giving them productive land). After all, the professed aim of R&R is to improve the standard of living of people displaced.

Unless we take such a view of R&R, it cannot succeed.

According to Datye, displacement itself is not the problem, and certainly not an insurmountable one. After all, at the time of Partition, thousands upon thousands were uprooted from their homes in what is now Pakistan. Today, many of those people are prosperous Indians, and those who settled in Punjab have made it an engine of growth for India. This happened because there was a firm political will at the time, Datye pointed out, to settle people and give them an opportunity to make their lives.

Thus to say that governments cannot handle displacement on a large scale is wrong.

Only, the same political will has never applied to displacement caused by large dams. Datye rattled off the names of those he is familiar with: Tapi, Ukai, Chambal, Krishna and others. The Tapi project, he said, left the poor and tribal areas of the river valley—Nandurbar district in Maharashtra, for example—'high and dry', while the water went to the comparatively prosperous stretch of southern Gujarat. Those tribal areas 'got only displacement'; its residents were left to fend for themselves.

According to Datye, R&R can only work if displaced people 'get a share of development benefits equal to what project beneficiaries in the command areas and cities get.' And because that does not happen, because that is not even a view that is taken of displacement and R&R, because this naturally turns R&R into such a hard problem, we have to be willing today to consider options for dam projects that will reduce displacement. Even if those run counter to such decisions as the NWDT Award, that set out the framework for these projects.

In this context, Datye believes the Sardar Sarovar dam's height need not exceed 110 metres—at which height Gujarat will not lose any of its water, but displacement will be substantially reduced. (Chapter 4 will have some more about this). In any case, there is sufficient local storage capacity for water in Gujarat that the Sardar Sarovar dam need not submerge such a large area. This only strengthens the argument to reduce displacement as far as possible.

This notion—of using local storage *in conjunction* with water from dams—is another pillar of Datye's model for how R&R must proceed. Once oustees have title to their land and access to water—including local sources—Datye believes that an integrated approach to using that land— part agricultural, part horticultural and part devoted to forest produce—will make it productive. Thus, an agenda for the reform of water, agriculture and allied sectors must be part of the plan for any large dam project. It must also be a part of alternate strategies, such as watershed development, that the NBA and others advocate.

Another angle to this is Datye's belief that recent advances in wind and solar energy—they are now viable energy options—make it unnecessary to store water for hydroelectric generation beyond November each year. This supports the case for lower and fewer dams. 'New knowledge and technology advancement,' he wrote to me after we met, 'calls for radical re-orientation of river valley development plans made 20 years ago.'

Without recognizing all this, said Datye, 'saying that we cannot reduce the height of the dam by even a metre is also unreasonable.'

•

To end this chapter, let's return for one more look at R&R. Let me remind you once more of its professed aim—and not just in India: to actually improve the standard of living of people displaced by 'development' projects. The World Bank makes a point of mentioning this in loan agreements, as does the NWDT Award, and the states' own R&R packages claim it as their goal. In fact, the NWDT Award puts it in some context, which is worth paying some attention to. Here's what it says:

> The displacement of the people due to major river valley projects has occurred in both developed and developing countries. In the past, there was no definite policy for rehabilitation of displaced persons associated with the river valley projects in India . . . [C]ash compensation [under the provisions of the Land Acquisition Act, 1894] was the practice, which resulted in the resettlement of displaced families becoming unsustainable due to squandering away of the compensation money. This type of rehabilitation programmes deprived the poor and illiterate tribals from their land, houses, wages, natural environment and their socio-economic and cultural milieu.

In other words, the NWDT makes a clear and frank indictment of the way R&R was treated before its time, and especially—let me emphasize that, *especially*—of the dangers of handing out cash compensation as the Land Acquisition Act dictated. Even loan agreements signed with the World Bank for Sardar Sarovar explicitly forbid cash compensation: 'In no case shall cash payments be made in substitution for actual rehabilitation.' (Morse: 30, quoting Development Credit Agreement signed by India, 1985)

Cash payouts, the NWDT asserts, only leave recipients destitute.

Mindful of the need to set this history right, at least for the future, the NWDT Award itself spells out a liberal package for displaced people (which, of course, was further liberalized by the state governments.) It also lists the 'main objectives of R&R'. The first of these is to 'improve the standard of living or at least regain the standard of living they had been enjoying prior to their displacement.'

It was with these NWDT Award guidelines and requirements in mind that I began reading a letter from an official in the Government of Madhya Pradesh to the Narmada Control Authority (NCA). Written in May 2001, the letter forms part of the agenda for the fiftieth meeting of the NCA's R&R subgroup, which I mentioned earlier. It reflects the Government of Madhya Pradesh's anxiety about the prospects of R&R it will have go through for people displaced by dams on the Narmada. The letter notes that the number of project-affected families (PAFs)

is increasing day by day necessitating requirement of larger area of land in Madhya Pradesh. . . . According to Government of MP, there is paucity of cultivable government land in MP. As pointed out at various fora, it is difficult to arrange more land for the oustees [who choose to stay in MP]. Although efforts are on to identify and procure as much land as possible for allotment to the oustees, *past experience does not raise hopes very high.* [Emphasis added]

Here is the government of the state that has the largest number of oustees, admitting to problems with R&R. It is precisely this 'past experience' that makes people question the dam; in particular, they did so in the Supreme Court (see Chapter 9), though without much success. For the fact is, that past experience does not raise *their* hopes very high either. So you wonder: how is it that the Court explained away this unease by simply holding up the liberal R&R packages on display? How did they not take this 'past experience' into account? Why is it that seven months after the Supreme Court judgement, we have a concerned state admitting to significant problems with R&R? Why did it not inform the Supreme Court while the case was being heard? Or were Madhya Pradesh's concerns ignored?

If you think I'm making a mountain of a hillock, remember that Madhya Pradesh has, by far, the largest number of people who will be displaced by the dam. Even given the large discrepancies in numbers I cited above, this fact has always been true. For example, in the 1998 *FACTS* table that lists a total of 40,727 PAPs, 33,014 are from Madhya Pradesh: over 80 per cent. Of those 33,014, only 13,936 were willing to move to Gujarat, leaving over 19,000 for Madhya Pradesh to deal with. So when this state admits to difficulties in finding land for these people, it effectively means the entire R&R process is in trouble.

And yet this is hardly the worst aspect of Madhya Pradesh's confession. The letter continues:

> As such there is need for considering suitable amendments in Sub-Clause IV(7) of Clause XI of the NWDT [Award], and introducing a provision on the option of the oustee to accept *compensation in full in one installment* to enable him to purchase land and settle down as per his choice. [Emphasis added]

(Sub-Clause IV(7) of Clause XI, titled 'Allotment of Agricultural Lands', specifies that displaced families 'shall be entitled to and be allotted irrigable land', a minimum of two hectares per family. That is, it is the starting point of the liberal R&R package.)

And the letter continues some more:

> [The] Hon'ble Chief Minister of MP ... had very explicitly observed that, if the PAFs are to be allotted agricultural land and arrangement of land is to be made by the State ... then it will be very difficult to adhere to the time frame for the resettlement of the PAFs as approved by the Narmada Control Authority.

If the PAFs are to actually get what has only been promised to them, in packages that are repeatedly called 'liberal', not only will that be hard to do, but it will also take an inordinate amount of time. Again, was this not known while the Supreme Court was deliberating? Was this awareness of the time frame a recent development?

But the letter goes on:

> [Therefore] the Government of MP proposes that following [clause be added to the NWDT Award]:

> Provided that the displaced family shall have the option of obtaining full compensation for settling down and purchasing land in a village of his choice ... An option once exercised shall be final and no claim for allotment of land can be made afterwards [sic].

The agenda for the fiftieth meeting takes note of this Madhya Pradesh proposal.

Consider what has happened here. Starting from voicing the dangers of cash compensation, as the NWDT Award itself does, moving through spelling out an explicit (and, let's not forget, 'liberal') package for R&R that begins with two hectares of irrigated land, we have arrived at a point when the state with the most PAPs now wants to return to giving those PAPs cash compensation. 'In full in one installment', no less. It cannot find and purchase the required land, so it suggests to the PAFs that *they* take the money, that *they* go look for land. And on Madhya Pradesh's suggestion, this policy of cash compensation is set to become part of the NWDT Award: an Award that, to repeat, itself warns against cash compensation.

What is this R&R all about, anyway?

The letter ends by pointing out that this proposed amendment is 'in the interest of the displaced family as it would enable him to settle down and purchase land in a village of his choice at the earliest.'

Not only has the Government of Madhya Pradesh proposed this amendment, it wants us to applaud its magnanimity and vision. It wants us to believe that the cash compensation is in the PAF's own interest. Which argument PAFs will, if they know about the NWDT's case *against* cash compensation—that it has tended to reduce such families to destitution—find somewhat hard to swallow.

But, of course, there is in this very same letter an explicit mention of whose interest the Government of Madhya Pradesh really has in mind:

> This [proposed policy] will also help complete the project as schedule [sic].

It's going to be 'very difficult to adhere to the time frame' for R&R, we hear. But the project itself must be completed 'as schedule'. Therefore, it's back to handing out cash.

Why is it that while R&R is not bound by a schedule, every effort must be made to complete the project on time? When we hear so much about how R&R is vital,

about how it must be completed before construction on
the dam proceeds, I find myself asking: why should we not
approach this the other way round? That is, complete
R&R 'as scheduled' even if that makes it 'very difficult to
adhere to the time frame' for the project. Is that not
possible?

I have a feeling that if R&R is really treated as the
priority it should be, the project itself will face fewer
difficulties. But the way things are, I have a sneaking
suspicion instead. It's not a dam they are selling here, it's
a line.

4

A Glimpse of the Alternatives

The old woman hurries down the gentle slope as our boat gets ready to leave. She waves to a young man who is with us. He waves back with a shy grin. Then, as the boat begins moving, we notice that she is weeping. Tears running down her cheeks, she keeps waving as we move off down the swollen river.

Her name is Khatri Vasave. She lives in a tiny village called Domkhedi, on the banks of the Narmada in northern Maharashtra. The man is called Anil Kumar. He is from Pathanpara in Kannur district, Kerala. After a few weeks here, he is returning home. In those weeks, Khatri has grown very fond of this tall young engineer. And her fondness has roots, I suspect, in a bulb. One measly, not particularly bright, bulb.

For in this little hamlet, Anil and his colleague, Madhu, accomplished something small but profound.

They came here about six weeks earlier. Surveying the area, they found a small stream gurgling in the hills a short way from the village. Enlisting the help of the villagers, they built a one-metre-high, four- or five-metre-long dam across the stream. From the reservoir the dam created, they laid a pipe, past trees and slopes, to a concrete tank halfway to the village. From this tank, they ran another pipe down a steep slope about thirty metres, to a little shed they built at the bottom of it. In the shed, they set up a small turbine they had brought from Kerala. Finally, they strung wires from the shed to some huts in Domkhedi.

A turn of the valve one Tuesday evening, and there it was. On India's fifty-third birthday, for the first time ever, an electric bulb glowed in Khatri Vasave's hut. A few others lit up some other huts. In the weeks they spent here, Anil and Madhu had given these villagers what fifty-three years—636 months, a dispiriting number of days—of India's government had not. Electricity.

No wonder Khatri weeps to see Anil leave.

Domkhedi is due to vanish under the waters of the Narmada as they rise behind the Sardar Sarovar dam, just a short distance downstream. In fact, even at about 83 metres, the dam's height when I visited, Domkhedi already does get almost totally submerged at the peak of the monsoon. It was here, during the rains of 1999, that dozens of villagers and NBA activists stayed put while the water rose to waist, then chest, then chin level; they stood in protest against the dam until a nervous state administration yanked them out and arrested them.

They had to. For Domkhedi's villagers are being drowned out of their homes and off their land to build this dam that will, they are told, supply drinking water and electricity to areas of Gujarat very far from here. At some point, I imagine the irony of all this must have struck these villagers: nobody ever cared to bring Domkhedi's residents drinking water and electricity. And yet their lives here are expended so others in Gujarat can enjoy those things.

No wonder, too, that the people of Domkhedi simply do not want Sardar Sarovar to be built.

One year, they made that point by standing in the rising Narmada. The next year, Domkhedi remained a focal point of the protests against the dam. But in that year, something had changed there. Some huts actually had electricity. It was just as good, just as potent, just as desirable as the stuff the dam is supposed to deliver some indeterminate number of years from now, some indeterminate number of kilometres from here. Only, this electricity was produced right here in the village, generated with the toil and sweat of the villagers.

In some ways, this year's protest was the loudest of all: the demonstration that despite the odds against them, they were willing and able to produce electricity for themselves. And that, at a cost of about Rs 15,000, with help from Anil and Madhu.

Of course, it's important to retain a sense of proportion here. While six huts in Domkhedi are grateful for this power, on the face of it the idea does not seem practical on a larger scale. The turbine that's up and running in the village is a small one—it can supply 300 watts of electricity at best, enough for ten houses lit with low-power bulbs. (Though the families use compact fluorescent lamps—CFLs—which consume the same power but produce more light than ordinary incandescent bulbs). For the time being, there is power only for three or four hours every evening. Without doubt, this is a small effort; a 'micro-hydel' project that Anil himself told us is better described as 'pico-hydel'.

Besides, it's not clear this effort can be replicated outside hilly areas such as Domkhedi. The turbine needs steadily flowing water, which is why a tank was built here, the water piped thirty metres downhill to the turbine.

That's not possible in the plains, so what happens in villages that are without electricity down there?

After hearing all this from Anil, we trek up a narrow path, balancing precariously on the steep side of a hill, to the site of this small dam. The first thing that strikes me is how tiny the stream is. Yes, there have been two poor monsoons in a row, no rain for days now. But even so, this is a mere trickle. There's hardly enough water flowing in it to fill my bottle quickly, I think, let alone enough to fill a tank and then light a bulb, and more.

But then something else strikes me. I might have given this trickle no more than a glance. But two young men from Kerala saw its potential and persuaded these villagers of it. Here in Domkhedi, a few ordinary Indians have made creative use of their own resources. Scanty resources, sure, but they have been exploited right here, by those who live right here. And isn't that what self-reliance is all about?

Seen like that, you realize this is no mere toy on display. Anil, Madhu and the villagers take it very seriously indeed. Six working bulbs where there were none before: that's all the sense of proportion I need.

Besides, Anil tells me they have installed other turbines like this one in various Kerala villages. One supplies as much as 4,000 watts (4 kW). He believes the total micro-hydel capacity in Kerala is 2,000 MW. Compare that to the electricity the Sardar Sarovar dam will generate—1,450 MW—and you start understanding the potential of these minuscule turbines. Anil wants to make a study in this part of the Narmada Valley to estimate the capacity here. He adds, China has about 10,000 such micro-hydel plants in operation. They even supply power to the national grid there.

Not a toy. Not at all.

What's more, this dam also gives Domkhedi drinking water. The women here used to have to go up and down the hills every day to bring water to their families; in the monsoons especially, the water from the Narmada itself is very muddy. But now a pipe brings clear stream water from that tank right into the village. In addition, the two Kerala engineers are building bunds to prevent soil erosion in the stream's watershed. The idea is to give the water more time to seep into the ground and recharge the groundwater. In one or two years, they expect that this recharging will lead to the stream flowing round the year. Of course, that will make the micro-hydel project more useful still.

And if all that's not enough of a statement against a gigantic dam, there is even a pedal-operated generator, designed by an engineering student, operating in Domkhedi. Pedalled for an hour, it charges a battery enough to power a couple of bulbs for four hours. Make that *eight* bulbs where there were none before.

No doubt what's going on in Domkhedi is small. But it is at least a demonstration of what is possible. Of what the alternatives are to major projects that cause destruction and displacement; of what are the options to waiting for apathetic governments to act, to provide.

And it seems to me—I believe it was on this trip to the Narmada Valley that I fully understood—that this is what those who have fought this dam so long are saying: let us decide. Let us have a say. Let us make our lives, our futures. Don't take that away from us.

No wonder Khatri Vasave is crying.

●

When I wrote this story in a column on *rediff.com* in late 2000, I got several responses. To my surprise, though in retrospect it should not have been, many ridiculed me for comparing the Sardar Sarovar project to a little turbine in Domkhedi. What happens to the millions across Gujarat who need water and electricity, they asked me. Can a little turbine, even many of them, supply their needs?

There are a few reasons I told the story then, and do so now. One is to make the point that the dam will bring benefits to much of Gujarat, if it ever does, years from now. This means that if people in Gujarat want solutions to their water and power problems, if they want them *now*, they have to look elsewhere for answers. That's what drives Lirabhai in Toraniya, for example. And here in Domkhedi, that's why they chose to try out this little turbine.

A second reason is the spirit I alluded to at the end of the story: that in some ways what the people of Domkhedi are really saying is, let us solve our own problems. Don't decide for us, don't presume to decide for us. Just grant us the chance to live as we want to. By their very nature, dams like Sardar Sarovar, the development they represent, ignore that spirit.

A third reason is that what happened in Domkhedi is an entirely different form of protest against the Sardar Sarovar dam. Not a dharna or a march or gathering to stand for hours in rising water, but a simple demonstration that neither people in Domkhedi nor anyone else need wait for that dam to be built.

And a fourth reason is that the Domkhedi micro-hydel project represents the essence of what critics of the dam have said for years: search for small, local alternatives that will benefit people right now. Not decades hence. When you search for such alternatives, the scale is hardly the point. Sure, one talao in Toraniya does not turn it into another Punjab; sure, six dim bulbs blinking in Domkhedi do not turn it into a New York City. But that's still six more than Domkhedi has ever had, going back to the times of Neanderthal Man.

Unfortunately, the preoccupation with the dam—from both sides of the argument—has meant that few people have studied the small alternatives. And yet they offer hope for Gujarat.

This is something that some people do recognize: in my experience, in Toraniya and Domkhedi.

Nor does this apply just to people in these villages. After all, every year we read about enormous water problems in towns such as Rajkot, where taps flows for only a couple of hours every few days and tankers must supply water to residents. Those more middle-class, urban citizens of Gujarat also need water right now.

They need alternatives to the project right now.

And that is why, even if the Supreme Court has allowed construction on the dam to proceed, a consideration of the alternatives to the dam remains as relevant as ever.

●

At other places, you will find more about possible alternatives to the SSP. Here, I will briefly describe only one of them, the so-called Paranjape-Joy proposal. I think it captures some of that spirit I have mentioned above.

Based in Pune, Suhas Paranjape, an engineer, and K.J. Joy, a social worker, have been interested in, and working on, water management issues for many years. In 1995, they wrote *Sustainable Technology: Making the Sardar Sarovar Project Viable.* This is a detailed proposal to restructure the SSP, starting from where it stood at the time. It is a great pity that this lucid book never got the attention it deserved, particularly from those who are building the dam. Nor is there any sign in the October 2000 Supreme Court judgement that the Paranjape-Joy proposal was given any consideration.

'We believe', wrote the authors in their introduction,

> that our proposal is capable of satisfying the legitimate
> interests and demands of each of the main parties to
> the long-standing dispute over the [Sardar Sarovar]
> project, and can thus help the current deadlock
> between the supporters and the critics of the project.
> (Paranjape and Joy: 5)

Quite a claim, especially if you know how polarized the 'main parties' have become over the years. What is this proposal that promises to bridge the gap between the 'no-dam'-*wallas* and the 'dam-at-all-costs' people? As its authors make clear, it

> concentrates on bringing about a substantial reduction
> in submergence and displacement, fully utilising the
> quantum of Narmada waters allocated to Gujarat,
> improving the equity of water distribution within
> Gujarat and ensuring sustainability in water use, and
> achieving a substantial *increase* in water and energy
> benefits accruing to the three major states involved in
> the project. [Emphasis in original] (Paranjape and
> Joy: 8)

Again, these are some pretty lofty claims. How will Paranjape and Joy succeed in accomplishing all this?

Naturally, it is not possible to explain all the book says in a few paragraphs. But consider a few aspects of their proposal. First, a look at the more detailed statement of objectives they hoped to achieve:

> a) To drastically bring down the displacement—
> especially the displacement of entire villages—to such
> an extent that the displaced can be rehabilitated

within the contiguous upstream area and within their own socio-cultural milieu;

b) To ensure that the 9 million acre feet (MAF) of Narmada water allocated to Gujarat by the Tribunal is not reduced, and that this water is made available on *a priority basis* to Kutch, Saurashtra and North Gujarat; [Emphasis added] (Paranjape and Joy: 10)

Point (a) would be accomplished by limiting the height of the dam to 107 metres (351 feet), compared to the currently planned height of 138.7 metres (455 feet). Paranjape and Joy calculate that this will reduce the area to be submerged—and therefore the displacement of people—from 39,000 hectares to 10,800 hectares. More interesting than this mere reduction was their proposal to rehabilitate PAPs in 'the contiguous upstream area': more about that below.

The second point is particularly interesting. As you have seen, the SSP has always been sold to Gujarat as the 'lifeline' of that state, and especially of perennially water-short Kutch and Saurashtra. This has been repeated so often that it has become something of an article of faith in Gujarat, stimulating sentiment for the dam that cuts across all political boundaries in that state. After all, rhetoric about a 'lifeline' is such a sure way to shore up political support that no party can choose to ignore it.

And yet the truth is that according to the project plans, these areas get only measly allocations of water (2 per cent to Kutch, 22 per cent to Saurashtra [Paranjape and Joy: 11]). What's more, the dam will deliver the water there only decades from now (2025 was the estimate *before* the Supreme Court case was filed). Why then is this project claimed to favour thirsty Kutch and Saurashtra most of all?

Silly question, of course. It is claimed as such purely

for political mileage. More intriguing is the related question: when all the evidence shows how low these two districts figure on the priority list for water from this dam, why has it come to be so widely believed that the dam is a lifeline for Kutch and Saurashtra?

Paranjape and Joy, like many others, are uncomfortable with how the project plans to treat the neediest parts of the state. Their proposal explains their determination to see that such a dam sends water first to these parts. In addition, their proposal also increases Kutch and Saurashtra's share of Narmada water. They recognize both the need for water in these areas, and the years spent tying the dam to the well-being of Kutch and Saurashtra. Doing so, they call the long-time bluff of the Gujarat government. The other objectives are:

c) To make the availability of this water conditional on the adoption, by the beneficiary villages, of regenerative, sustainable and equitable water use, thus opening up a path to self-sustaining prosperity; and

d) while achieving all this, to salvage as much as possible of the construction and expenditure that have already taken place on the project, by incorporating it in the alternative system and thus reducing wastage as far as possible.
(Paranjape and Joy: 10)

Objective (c) is novel. Has there been any other dam proposal that sets out such a condition? Unlikely, because water from dams all over the world has traditionally been supplied cheaply and in abundance. So much so, that it is almost expected that such water will never be rationed. What must we make of a proposal that requires beneficiaries to be—in a word—careful with water?

And finally, objective (d) takes care of a certain strain of thought, certain voices that work to stymie any debate. Pronouncing that such a lot of money has already been spent on its construction, they argue that the dam must perforce be completed by spending even more money. Otherwise all that money spent will go waste. It's a dubious argument at best, and Paranjape and Joy are careful to head it off right at the beginning.

With these objectives, the proposal shows that without any increase in cost, it is possible to:

• build the dam to a lower height, thus reducing submergence and displacement greatly;
• give Gujarat the same quantity of water from the Narmada;
• more than double the area served by the project;
• substantially increase the water shares of Kutch, Saurashtra and North Gujarat;
• and more.

One element of the proposal is critical to the aims Paranjape and Joy have in mind. That is a profound change in how we view water coming from a dam. The conventional view of irrigation is that dams are independent water delivery systems, whose operation need not take note of local conditions in the areas to where they take water. Paranjape and Joy argue that this *exogenous* water from dams is really best used to refresh groundwater and local surface storage systems. In fact, they argue that such water must be provided

> only after local sources and systems have been adequately strengthened and developed, and on the basis of such strengthening. [Emphasis in original] (Paranjape and Joy: 27)

In other words, a local farmer, served by water from a dam, must see his water as coming to him from various different sources, only one of which is that distant dam. Most of those sources lie around and under him, and the water from the distant dam actually helps replenish these local sources before it becomes available to him directly. That is, the water he uses is a combination of exogenous and *endogenous* water.

Another interesting feature of the proposal is the responsibility it places on the people who will get water from the Narmada. Unlike in other dam projects, it does not stop at delivering water and doing it cheaply. It demands that certain measures be taken by the users of the water as a *condition* for getting the water.

For example, Paranjape and Joy propose that beneficiary farmers adopt an alternative cropping pattern to make best use of the water. About one third of the water would be devoted to food grains; another third to agro-forestry; the last third will be used for 'marketable produce' like fruits and vegetables. (Recall Datye's model that I touched on earlier). This is to encourage the beneficiaries to give over part of their land to what the proposal calls 'permanent vegetative cover.' In fact, the proposal says:

> We have therefore included the provision of
> permanent vegetative cover on 1/3rd of the service
> area as one of the minimum conditions to be accepted
> by the water users' group in return for the provision
> of basic service. (Paranjape and Joy: 96)

Which other irrigation scheme makes such a demand of the beneficiaries, places such a responsibility on them? Think of the value the users will place on the water when it comes to them not at an absurdly low cost, but with a condition like this attached.

Think, in fact, of the value of making users aware of the *responsibility* that using water places on them.

As novel as this approach to water use in the command areas of the dam is Paranjape and Joy's thinking about rehabilitation. To begin with, they call it 'the most important issue' (p. 103) in the project, those words themselves a departure from more conventional plans. (Elsewhere, they make it even more explicit: 'The process of rehabilitation needs to be given priority if the dam is to be completed at all.' [p. 161]) Then they propose a lower height for the dam, greatly reducing submergence. And as I noted above, they then propose to rehabilitate the oustees *upstream.* As they write:

> [P]urely on the ground of social justice ... it is necessary for every project to make available a definite quantity of water for the upstream areas of the dams, which, after all, bear the brunt of the human costs of the project. ... If [this is done], then rehabilitation in upstream areas becomes possible without having to uproot people entirely from [those] areas with which they are familiar, and where they have their economic, social and cultural ties. These avenues have not been explored at all in the Sardar Sarovar project.
> (Paranjape and Joy: 104-05)

Paranjape and Joy set out an 'influence zone' for the area of submergence, contiguous with it. The greater part of this zone will be in Madhya Pradesh, since most of the submergence will happen there. It will cover about 100,000 hectares. The same model of water use that they propose for the downstream command area will apply to this zone as well. Thus the loss of forest land will be more than compensated—in area, if not in actual ecological value—by the creation of permanent vegetative cover in this upstream influence area.

But the really novel aspect of this proposal is the way the zone will be used for R&R. Of the 100,000 hectares, Paranjape and Joy estimate they will use about 10,000 hectares for R&R. They believe no more than 5 per cent of the 'lowest grade forests' in the three districts of Jhabua, Dhar and Dhule will be lost to provide this. Dam oustees will be resettled and rehabilitated in this area. R&R becomes 'part of [the zone's] overall development'; which development, of course, will use the water that the project makes available upstream.

This hints at the reason for this idea of an influence zone: this is a view of R&R not in isolation, but as an important part of an integrated plan of development.

There is much detail in this R&R plan that I can hardly spell out here. But three final aspects are of interest.

One, they estimate that the 'definite quantity of water' needed for this influence zone amounts to just 0.1 MAF, an insignificant fraction of Madhya Pradesh's allocation of Narmada water.

Two, they will charge the costs of pumping this much water upstream, to Gujarat. This is because Gujarat is the main beneficiary of the project, and the main beneficiary too of the reduced submergence and this novel way of tackling R&R.

Three, and I can do no better than quote them again:

[B]y providing basic service [of water] for the entire influence area, we would be stabilising the livelihoods of the adivasis [there]. It is often not sufficiently realised that the adivasis' overexploitation of the forests is primarily due to the increasingly precarious nature of their livelihoods. . . . In fact, ensuring stable livelihoods for the adivasis is the most important forest protection measure, and also the most neglected. (Paranjape and Joy: 108)

Which other plan for a dam makes, or has made, explicit provision for some of our country's most forgotten people, and does so by first taking their own situation into account?

•

In this short look at the Paranjape-Joy proposal, I wanted above all to give you a flavour of the concerns that fuel it. I mean concerns such as:

• the need to address, first of all, the suffering of the most water-deprived residents of Gujarat;
• the need to make water users aware of, and responsible for, how they use water;
• the need to reduce the enormous human cost of displacement (what they actually call the 'greatest cost the project has to bear');
• the importance of making R&R a priority;
• and the importance of giving adivasis a dignified life.

Striking about this proposal is its profound awareness of the humans who are part of the process of dam-building, of development itself. Nowhere in this entire proposal did I find claims that the dam would use 'the highest quantity of concrete to be placed in any dam in India'; or that the canal 'will be largest capacity irrigation canal in the world'; or some apparatus to be used in the construction 'will be nearly 1½ times the height of the famous Kutub Minar.' Whereas these feature prominently in the SSNNL literature.

In other words, the Paranjape-Joy proposal represents an utterly new view of development: that its success must be measured—can only be measured—in terms of its effect on the lives it touches.

For that reason, even if this proposal is ignored, or somehow determined to be unfeasible, the outlook it represents must be preserved for future projects. That is the greater sense in which it is an alternative.

5

Displacement: Misery Compounded

Parvat Varma's boat has a neatly painted slogan on its prow. '*Mera Bharat Mahan*', it says. My English companion turned to ask me what it meant. After I explained, I looked up at Parvat, standing at my shoulder. 'So Parvat, tell me,' I said. 'Is your Bharat *mahan*?'

Parvat stopped poling the boat and thought for a moment. Then he pointed to the two large piles of sand that lay beside us. 'See the colour of that sand?' he asked. 'It's like gold, isn't it? Is there any other country in the world where I can get gold like this? Of course *mera Bharat mahan hai!*'

It was one of those moments I experience every now and again, when something said simply overwhelms my cynicism about the country I live in. So I nodded weakly and tried to force down the sudden lump Parvat's ordinary faith produced in my throat.

Parvat and his boat-mates, and several dozen others like them in the village of Pathrad, get down to the riverside by 7.30 every morning. Throwing their two shallow pans into the boat, they then pole it about a kilometre upstream. There, they drop anchor and get to work. 'Work' means leaping into the water, diving under and digging sand from the bottom. Over and over. One pan at a time, they lift the sand and pile it into the boat, making it settle steadily lower in the river, until its side is nearly level with the water: several hundred kilos of sand, I imagine. This takes them about forty-five minutes or an hour of steady, drenching toil. When it is full, they pole the boat back to a place on the bank near the village, where the women, kids and a few other men wait. Parvat and mates lift the sand, one pan at a time again, on to the heads of these helpers; in turn, they wade ashore and empty the pans there. Large piles of sand come up; simultaneously, as the boat empties, it rises imperceptibly in the water. Another thirty minutes later, Parvat and mates throw the pans in the boat and set off upstream again.

They run through this cycle about six times every day. By about three in the afternoon, they are exhausted and ready to stop.

The return journeys with a full boat are always precarious: 'if there's a wind,' says Parvat, 'the boat sometimes goes under.'

'What do you do then?' I ask.

Silly question. 'We pull it out and fill it again,' says Parvat, smiling at this display of city-bred naïvetè.

But the wind sometimes helps, too. It means these men can use a sail and put in less effort to pole the boat. Then again, some boats don't have sails. In them, one man goes up to the front, takes off his dhoti, pins down one end with his feet and holds up the other. Voila! A sail. There is something amusing, almost whimsical, about watching one of these boats scud past, powered by the wind in a dhoti-sail held down by ten sandy toes.

And what's it all for, you ask? Well, river sand is an important ingredient in construction material for city buildings. As Parvat and friends go about their diving and digging and shovelling, a steady stream of trucks comes to the bank, ready to be loaded with the sandy fruit of Parvat's labours. They deliver the sand to construction sites as far away as Indore, a five-hour journey from Pathrad. One truck driver tells me they occasionally even take a load to Bombay, though in general that's too far to make it worth the trip.

So it goes, here on the banks of the Narmada, upstream from Maheshwar in Madhya Pradesh. If this gorgeous river is revered by millions, the hum of sand-quarrying near Pathrad is one reason. After all, Parvat tells me, there's enough sand, enough work, to give all these workers a daily take-home of between Rs 150 and Rs 200. That's no small potatoes. These men and women live in flimsy huts in one section of Pathrad, yes. Many of them are illiterate, yes. But they are by no means poor. They do a hard day's work with dignity and good cheer, and earn a reasonable amount for it.

Why is it, then, that I cannot shake one thought throughout the morning I spend with Parvat? The building where I live in Bombay, like those in cities all over the country, is built on the backs of men like Parvat. Quite literally so. For it is just this simple: if men like him are not willing to dive under the surface of a river to gather fine river sand, one pan at a time, for eight hours a day, day after day, how would we get this stuff? How much more expensive would our homes be?

Classic left-wing rhetoric? The rich making their way on the backs of the poor? You may think so, but at least three niggling details make it difficult to dismiss Parvat's situation so tritely.

One, Parvat and friends are not exactly poor, in the way we think of it in India. Two, this quarrying of sand has been happening in Pathrad for only about ten or twelve years. Three, there's a real chance that sometime in the near future, it will no longer be possible to quarry sand in and around the area.

The reasons for details numbers two and three involve two dams on the Narmada.

Pathrad village is divided in two parts. A straggly stream runs between them today. But once upon a time, only one side of the stream was settled. One year in the 1960s, because of a particularly heavy monsoon, the river rose threateningly. Much of Pathrad was flooded and many villagers lost their homes. A number of them then moved to higher ground, further from the river, across the stream, and rebuilt their homes there.

Today, this area of the village has *pucca*, well-built houses, a school, a few shops and a temple. In other words, it has an almost prosperous air about it. However, some people have remained in old Pathrad. They still live in mud and thatch homes that fit a more conventional picture of rural India. Scrawny dogs and chickens run about and there's a generally grimy feel to the place. Coincidentally or otherwise, the stream also marks a caste line in the village. New Pathrad has several families of Patidars, the farming caste that is widespread in this part of Madhya Pradesh. Old Pathrad has Kevats, fisherfolk, and Kahars, the sand quarriers we met. Neither of these is a traditional farming or landowning community. They make their living from the river. As Ghisa Lal, the *chai*-stall owner who caters to the sand-quarrying workers, told me: 'the river is our *kheti* [field].'

The Kevats and Kahars remained in the old village because at the time they could not afford to move. But more important, old Pathrad is close to the river, the source of their livelihood. And again, while their houses and surroundings are not quite as salubrious as in the newer section of the village, that livelihood does bring them a reasonable income.

Parvat Varma, of course, is a Kahar.

The really interesting thing about the Kahars of Pathrad is that sand-quarrying is not their traditional occupation either. They have been doing it only since about the end of the 1980s. Before that, they raised fruits and vegetables in the river bed: watermelons in particular, which are especially tasty when grown this way. (When I bought a watermelon at the market in nearby Mandleshwar and took it back to the home where I was staying, a man took one taste of the fruit, wrinkled his nose and said: 'This was not grown on the river bed.') In other villages in the area, the Kahars still cultivate fruit. And going by Parvat's and his friends' memories of their parents' lives, the watermelons and other crops brought in a reasonable income as well. Because it did, they stayed behind in old Pathrad when their neighbours moved further away. Close to the river, willing to risk the floods, still raising their fruits. After all, one year of floods did not mean they had to give up cultivating the fruit.

So why did they give it up in about 1990?

According to Parvat and his friends, the answer lies astride the Narmada, a few hundred kilometres upstream from Pathrad, some distance south of Jabalpur. It's called the Bargi dam. Completed as the 1980s wound down, this giant wall of concrete was the first large dam to be completed in the ambitious programme to 'develop' the Narmada Valley. And it is large indeed. I once chugged over the reservoir it impounds in a fairly fast little motorboat. There were times when we couldn't see either bank. All we could see was water all around us. Bargi's reservoir is an enormous and placid lake.

Bargi is also the first dam you run into, if you travel upstream from Pathrad.

When Bargi was completed, the pattern of flow of water in the Narmada changed completely. Upstream from it, of course, the river became that lake. But downstream, the flow was changed by the way water was released through the dam. How exactly that happened is not really relevant in Pathrad: the fact that it did is. The changed flow of water meant that the fruit and vegetable fields belonging to the Kahars became useless. All of a sudden in the late 1980s, they found that raising those tasty watermelons in the river bed was no longer a viable way to live.

That marked a turning point in Kahar lives here. At least in Pathrad, they were forced to give up cultivating fruit and to begin lifting sand out of the river. Today, the Kahars are now generally known as *reti-wale*. Sand-quarriers.

This switch from watermelons to sand is an interesting tale by itself. But there's a looming twist in that tale. If it comes to pass, it might just mean an end to the Kahars' sand-quarrying. Parvat Varma is fully aware of that prospect.

The reason for this, he tells me, lies astride the Narmada too. But this one is a few kilometres downstream from where he lives. It's called the Maheshwar dam. It's the first dam you run into, if you travel downstream from Pathrad.

The Maheshwar dam is not nearly as big and imposing as Bargi. For various reasons, there has been fierce opposition to it in the sixty-one villages it threatens to submerge—Pathrad among them—and that may explain the desultory fashion in which construction on it is proceeding today. Here is a paragraph from a June 2000 report on the dam:

> Construction work is limited to date. A causeway for construction vehicles was built across the river ten years ago. More recently, excavation was undertaken for the powerhouse, and some protective walls built to prevent intrusion of monsoon-fed waters into the site. After initiating this work, the developer and the site have been a focus of protests from people in the area who object to the R&R plan or do not want the project at all. As a result, work has failed to meet any of its deadlines, and an attempt to resume the project would require establishing new time lines for R&R and dam construction. (Bissell: 5)

The situation is more than mere observations in a report. When I went near the dam in a fishing boat, and later when I drove across it on a motorcycle, there were only a few workers pottering about; only a couple of trucks driving up and down somewhat aimlessly. It doesn't look like the dam will be complete any time soon. Or, in fact, at all.

But if the Maheshwar dam *is* finished, its reservoir will indeed drown Pathrad. Both sections this time, and for good. In particular, it will mean Parvat Varma will no longer be able to do his sand-quarrying. Not just because he will be driven from his home, but because at the place in the river where he dredges up his sand, the river will be too deep for him and his friends to dive in as they do now. Their own *kheti*, in a sort of ultimate irony, will rise up to devour them.

Simple, don't you think? One dam drove the Kahars out of the fruit-raising business. Fortunately, that didn't mean destitution: they began quarrying sand. A decade later, another dam threatens to drive them out of the sand-quarrying business.

'What does this mean?' I ask Parvat and several others on that river bank near Pathrad. 'What will you do for a living if that dam gets built?'

Amused, if resigned, shrugs are what I get in response. 'What will we do? Nothing,' they say. Older and wiser, Ghisa Lal has a sharper vision of the future. 'We'll turn into beggars,' he told me, 'and stick our hands out in some big city. What else?'

•

I have read extensively about the Narmada dam projects; travelled in the Narmada Valley a reasonable amount; but I think it is fair to say that few experiences were quite as telling as the tale I have related above. Nothing captured the futility and cruelty of the way we pursue development, as that day spent with the sand-quarriers in Pathrad. It comes down to some simple questions: why must these people have their very livelihood threatened not once, but twice? Why should they have to assume that dams mean they will soon be reduced to begging?

And because they must, how should the people of this Valley react to the destruction they face from the dams that are planned for the Narmada and fast-flowing tributaries?

One morning, on another dam, I found myself smack in the middle of an answer to that question.

•

By all rights that morning, I should have been dead to the world. We had spent four hours, till midnight, being flung about in the back of a truck designed to magnify every bump on the road into a mountain. Another hour was spent in the back of a jeep where I felt the bumps slightly less, but whose driver must have been deaf, going by the volume at which he played a series of screeching and out-of-tune songs. I was exhausted when we reached the village of Khedi-Balwadi and flopped on to a too-small, but welcome, string-cot to grab some sleep.

I should have slept like a log. But no. There was too much activity. Sometime around four in the morning, I finally gave up trying and lay there awake, listening to the sounds of the dozens of voices around me, savouring the barely suppressed air of excitement. These people were about to make a statement, and the thrill was evident.

People were gathering in the little village from various surrounding hamlets. I myself had come with a cheerful group from Pathrad, over a hundred kilometres away. By the time the night sky began fading, hundreds of men, women and children packed the spaces between the huts, chattering and laughing and waving little flags. I got up, washed my face and checked that I had film in my camera. Then we were off.

Singing songs and raising slogans, high-spirited and charged with enthusiasm, our procession fairly raced through fields, over low rises, over the rocks of a dry stream bed, over a couple more rises—and suddenly we were at our destination. A large stone and earthen wall, a tall crane hanging over it, a concrete spillway, and another tall wall on the other side of the spillway. This was the site of the dam being built on the Maan river, a tributary of the Narmada, in the south-western corner of Madhya Pradesh. The dam that threatens the existence of Khedi-Balwadi and sixteen nearby villages.

Without so much as a pause, slogans still ringing in the clear morning air, the procession swarmed up two bamboo ramps and on to the spillway. Within fifteen minutes, flags and banners were strung up all around the dam. To much cheering, one young man climbed the crane, finally heeding the shouts that warned him about a beehive halfway up, and unfurled a large flag on top. From the time we had set off from Khedi-Balwadi, it had been no more than thirty or forty-five minutes. These few hundred villagers had done what they came here for: to 'capture' the Maan dam as a protest against the destruction it would bring to their lives.

This sunny day, there would be no work on this dam site.

A couple of hours later, this message was underlined. Two orange dump trucks rumbled down the steep slope below the dam. Filled with construction material, they were headed for an open area to dump the stuff. Without prompting, a group of fifty or so women poured off the dam and raced after the trucks. The rest of us watched, spellbound, from atop the dam. '*Nari shakti, Narmada shakti*' ('Woman power, Narmada power'), we could hear them shouting. They caught up with the trucks just as they were about to unload their material. A short confrontation ensued, and suddenly the two trucks turned tail and sped back the way they came. Only, they could not get back up the slope with their loads. One finally deposited half its material at a turn in the road and only then was able to wheeze slowly up the hill. The other waited for an excavator that arrived some time later and pushed it up.

From on top of the dam, from the women far below, a joyful cheer went up. Yes sir: no work on the dam today.

Much later, a police team turned up, rounded up many protestors—fifty-two children among them—and took them to jail in nearby Dhar. As is now routine at such times, the police managed in the process to rough up several of the villagers. The police claimed the villagers threw stones at them while being arrested. More impressive, they charged the villagers with—yes—atrocities against tribals.

I suppose that while they were being bundled into vans and carted off to jail, these women and children and men took time out to identify, locate and then beat the living daylights out of a few random tribals. Right.

●

Let me be clear here: it's not as if the protestors did not expect to be arrested. Or to be accused of all manner of things. They did. I am not trying to draw your attention to the injustice of the arrests, nor to the strange charges against these people. In some ways, all those things are mere details, just the sideshow. What I found myself thinking about up on that Maan dam were a few random questions.

Like: what drives several hundred people, including children and women nursing babies, to race pell-mell on to a dam early one morning and sit there, knowing full well they are going to be arrested? What makes a young man climb precariously up a crane? What sends a few dozen of the women running after two orange dump trucks? Really, what motivates ordinary rural folks to take a stand against all those cherished notions of 'progress' and 'development', to confront the might of the state in doing so? Would you or I do the same?

Actually, I think I know why people protest like this. Through the days before the storming of the Maan dam, I had wandered in some villages—like Pathrad—just upstream of the Maheshwar dam. These villages will drown if that dam is completed. Over those few days, and that morning both in Khedi-Balwadi and atop the dam, I spoke to many people about these two river projects. The NBA is active in the area, and, in fact, organized that protest at the Maan dam. Very deliberately, so I could also meet people who might disagree with the NBA—and not even the NBA would claim everyone in the area supports its fight—I had chosen to wander about alone.

Sure enough, there were people who believed the dams should, and would, be built. There were even some who thought they would bring good both to them and to the country: the 'progress' of the nation and all that, right? But from them, as well as from the rest, I heard one thing again and again, and it simply amazed me.

These dams have been under construction, therefore threatening to push people from their homes, for several years. Yet everyone I spoke to said that not a single person from the government had ever come to tell them what the dam would do to their homes, or about any rehabilitation programme. Never. Some of these people were actually quite willing to accept whatever package the government offered, and move: providing they were given such a package in the first place. But that had not happened.

In fact, the first they heard about the Maheshwar dam—not a rehabilitation policy, note, but the dam itself—was from NBA activists who came to their villages. (Unless you count watching the construction begin as they went about their lives.) It was only then that they realized the dam plans included driving them off their land and homes.

Up on that Maan dam, I tried to put all this in some perspective. Now I live in the Bombay suburb of Bandra. Let's say that one morning, I look out of my fourth-floor window and see construction starting up on a dam a mile away, in Khar. For some years, I watch the Khar dam rise, aware that such a huge project could only have been conceived by the government. But in all those years, nobody from the government comes to me, nor to any of my neighbours, to tell us what the consequences of building this dam will be for our homes.

When I find out that one of those consequences just happens to be that my home will end up under water, how do you think I might react? How would my neighbours, my fellow citizens of Bandra, react?

How would you react?

Do you think you might consider swarming on to that dam? Maybe even running after a dump truck?

•

Somewhere in my papers, I have a copy of a letter from the Government of Madhya Pradesh's Narmada Valley Development Department to the vice-chairman of the Narmada Valley Development Authority in Bhopal. Dated 2 May 1999, it lays out seven different 'orders' about the implementation of various dam projects in the Narmada Valley. One of those orders reads:

> After reviewing the current situation in the Maan and Jobat [another Narmada tributary in MP] projects, it has been decided that ... a Rehabilitation Planning Committee will be constituted. ... The Committee will keep in mind [that] the rehabilitation and resettlement of families living in the areas likely to be submerged due to the construction work up to 15th June of any year, should be completed by the 31st of December of the preceding year as per policy.

That morning on the Maan dam, I roamed among about 400 people 'living in the areas likely to be submerged due to the construction work up to 15th June of' the year 2001. They were 400 out of some 5,000 people—993 families in 17 villages—who were likely to be so submerged in 2001.

'As per policy', at least as I understand that letter, their rehabilitation and resettlement 'should' have been 'completed by the 31st of December of the preceding year': the year 2000.

Except, it wasn't. Now that's their very personal experience of this 'progress' and 'development'.

Well, that and being charged with atrocities against tribals.

•

Perhaps it takes meeting a Parvat Varma, or comprehending what pushes people to 'capture' a dam, to understand what R&R has really meant in Indian projects. That poor record alone should make us pause in our dam-building efforts. Only, it does not.

I suspect the reason is that many of us believe implicitly in a most appealing little equation. If the country is to 'progress' and 'develop', it goes, 'some people' will have to 'sacrifice'. 'Some people' must pay a price so that many others—many *more* others, we can presume—can 'benefit'.

It's a seductive argument. But fifty-five years after independence, it is about time we began to examine it.

For one thing, it is not clear that the country has in fact 'progressed' and 'developed'—to the promised extent, at least—because of the myriad dams we have built. Many millions have sacrificed homes, lands and lives to those dams, but India has more absolutely poor people than we had Indians in 1947. The very electricity those dams generate still bypasses thousands of villages; it might be an Indian irony that some of those villages are within shouting distance of some of those dams.

For a second, plans to build a dam never include a clause saying the people who will be worst affected must be the first to benefit from it. (Recall the Paranjape-Joy proposal). No doubt such a policy might cause logistical problems; but at least giving them priority, as an *intrinsic* element of dam-building, might ensure that they end up better off than they were before the dam. Instead, they usually end up much worse off.

For a third, the set of people who 'sacrifice' and the set of people who pronounce that 'some people' will have to 'sacrifice' are, as mathematicians would say, disjoint sets. In my experience, nobody belongs to both sets. That is, those who speak about sacrificing never, in my experience, have to sacrifice anything themselves.

Fourth, the same people sacrifice, every time. Over and over again in some cases, as with Parvat Varma.

The truth about India's dams is that the people they have uprooted have never been treated well. What's worse, authorities responsible for the dams routinely deny this, paint rosy pictures of the way R&R has progressed. For example, take these excerpts from the Supreme Court majority judgement of October 2001:

> If one compares the living conditions of the PAFs in their submerging villages with the rehabilitation packages first provided by the Tribunal's Award and then liberalised by the [three] States, it is obvious that the PAFs had gained substantially after their resettlement.

> [T]his Court is satisfied that more than adequate steps are being taken by the State of Gujarat ... and therefore, continued monitoring ... may not be necessary.

As you will see in chapter 9, the Judges were satisfied by a mere exposition of the *package* of R&R the state governments had put together. They were content with an affidavit from the State of Gujarat reporting its 'adequate steps.' It did not concern them to find out how closely these statements matched the actual situation of people affected by the projects.

Dams displace people, inevitably. It's how the builders of dams treat those people—the progress of R&R, in other words—that worries me and many others. The 1991-92 Morse Report did some investigation into R&R. Here are some remarks from it about Gujarat's experience:

> In 1987-88 Gujarat developed a policy for its Sardar Sarovar oustees that has since been welcomed as among the most progressive packages of measures ever devised for securing the long-term rehabilitation displaced by large-scale development projects. Many told us in the course of our review that the Gujarat policies should become Indian, if not worldwide, norms by which resettlement policies should be guided.
>
> By comparison with what has been done in all too many other development projects, in India and elsewhere, Gujarat may be providing a basis for rehabilitation that represents a real advance.
> (Morse: 82)

Note how Morse begins by acknowledging the worth of the R&R package devised by Gujarat, just as the Supreme Court Justices did in their decision. However, the Morse Report goes beyond the terms of the package, to find out how it is implemented. 'Our consideration of Gujarat's implementation of R&R policy,' Morse writes, 'took us to many villages and sites, and we spoke with people from many places we were not able to visit.' (Morse: 88)

And the picture that emerged out of all that first-hand investigation is somewhat different from what the Government of Gujarat might claim. For as Morse points out:

> In [Gujarat], relocation is recent, resettlement is
> beginning, but rehabilitation is for the most part a set
> of promises about the future. (Morse: 83)

Controversies about the plight of people displaced by the
SSP date back to 1961. In that year, land was acquired in
six villages around Kevadia, to build roads, offices and
accommodation for the engineers working on the project.
The Kevadia Colony, as it is called, is tidy and beautifully
landscaped, and is the project's nerve centre.

Estimates are that about eight or nine hundred families,
over four thousand people, were affected in those six
villages. They told Morse they had lost up to 70 per cent
of their land. Forty years later, they are still waiting for
reasonable compensation. Here's how Morse describes
what they told the Committee:

> Many villagers told us that they had been given
> between Rs 90 and Rs 250 per acre. ... Still others
> told us that they had, at the time and subsequently,
> been promised various compensation packages,
> including secure employment at the Sardar Sarovar
> site, patches of alternative land, and social benefits at
> their villages. ... Resentment has built up among
> many of the people of these villages about the low
> level of compensation. ... They felt cheated.
> (Morse: 90)

> [S]enior officials in Gujarat told us that retrospective
> application of their 1988 policy [on R&R] would
> create a potentially umanageable precedent: people
> affected by other projects all over the state might
> thus be able to seek similar benefits. ... The Kevadia
> villagers' demand for oustee or project-affected person
> status has been persistently rejected. (Morse: 91)

> Both non-government organizations and World Bank representatives have strongly criticized Gujarat's failure to deal fairly with the six villages. To refer to problems of precedent (which can be solved by government action) ... looks like evasion of the issue.
> (Morse: 93-94)

> At [the time of a 1985 World Bank Report] Gujarat had ignored the plight of the six villages for over 20 years. Seven more years have now elapsed. (Morse: 94)

Need I add: ten more years have now elapsed. Only a few months before I wrote these words, I actually met a man on the banks of the swollen Narmada who was pushed out of Kevadia. He still hopes to be compensated. Is this the 'sacrifice' we have in mind as we glibly use that word?

Morse examined other specific aspects of R&R in Gujarat in some detail. Here are some points from the conclusions they make:

> In the early stages of resettlement ... provision of basic resources and facilities left much to be desired. ... After 1988, when Gujarat completed extension of its resettlement policy, the provision of land continued to be beset with difficulties. ...

> [T]he policies [have] never been designed with the particular social, economic and cultural needs of the oustees in mind. Hence the hardships caused to oustees. ...

> Gujarat has indicated that it is prepared to provide R&R to all the oustees from Maharashtra and Madhya Pradesh who wish to take advantage of Gujarat's policy measures. ... There is little basis for concluding that Gujarat would be able to achieve such a task. ...

> [T]he core problem [is that] R&R have been effected pari passu [with dam construction].

The provision of irrigated land to the more entrepreneurially minded or market-oriented oustee families could secure a long-term economic basis for their lives. This group seems to us, however, to constitute a minority of the total oustee population in Gujarat. When it comes to the prospects of others affected by the Projects . . . serious doubts must arise. . . . [S]erious misgivings must remain. (Morse: 132-34)

Do you believe that any of all this has changed, and for the better, since 1992? Yet, as you will see, the Supreme Court wants us to think that it has.

• • •

But let's assume, as the Government of India did, that Morse and his report are not worth paying attention to. Even so, there have been many other signs of the way governments look at R&R. In 1993, I met someone who had been 'resettled'. His name was Keshuram Dhedya.

In March that year, I visited Manibeli, a little village on the Narmada not far upstream from the site of the Sardar Sarovar dam. Without being spectacular, there was a kind of heartfelt beauty to this little village. There were hills all around; a massive banyan tree shaded the little village square; at the bottom of the gentle slope that started from the furthest-flung branches of the tree, the Narmada river flowed past languidly. I slid down the slope to bathe in the river. As I sank into the water, the loveliness of this tiny hamlet hit me—but like a fist in the stomach.

For I knew very well that this spot would soon vanish under a sheet of water. The Narmada, as dammed by the Sardar Sarovar dam, would soon submerge Manibeli.

One afternoon in Manibeli, I spent some time with Keshuram Dhedya. I remember chatting with him. I remember him introducing me to his wife in their little hut, a few dozen yards upriver from the banyan tree. I even remember the hot cup of *chai* he made for me. So it was with considerable surprise that I recognized his hut in a magazine picture, just a few days after I returned from Manibeli. The picture featured in an article that extolled the measures authorities were taking to rehabilitate those whose houses were going to be lost to the rising river. Like the residents of Manibeli.

The caption below the photograph informed me that this was the 'last hut in Manibeli' to have been shifted to higher ground, a hut belonging to Keshuram Dhedya. True enough, it did belong to Dhedya. But I had been inside that very hut only days earlier, much later than the date of this issue of the magazine. It had not moved anywhere, let alone to higher ground. It was decidedly where it had always been. As was Dhedya himself.

I returned to Manibeli in June that year. This time, a huge expanse of water had submerged everything, even the banyan tree. There was something irretrievably sad about the sterile panorama before us. It was hard to imagine this vista as the lively, pretty spot that I had visited just three months earlier.

Then I saw a little hut, still where it had been in March, still with the same people living in it, though now it stood just above the water level. Keshuram Dhedya's hut. Strangely, the irony of it all left me happy—and depressed too.

1993 was not long after I first began taking an interest in the issue of dams on the Narmada. Call me naïve, but this small incident first opened my eyes to the way governments operate. Till then, I had no particular reason to believe or disbelieve much of what was said about R&R, whether by proponents and opponents of the dams. But

Dhedya's hut taught me a lesson—when it comes to official claims, scepticism is by far the best policy.

•

Perhaps you think all this is ancient history. Perhaps you think R&R issues are better treated today. Are they, really?

In early 2001, the Government of Maharashtra set up a 'Committee to Assist the Resettlement and Rehabilitation of the Sardar Sarovar Project-Affected Persons.' Headed by Shiraz M. Daud, a retired justice of the Bombay High Court, the Committee also had on it a Member of Parliament (MP), a retired tahsildar, a lawyer and two officials from the Revenue and Forest Department (R&FD), including its principal secretary. There were three special invitees: an activist and two joint secretaries, from the Irrigation Department and from the R&FD in the Government of Maharashtra.

In April and May 2001, the Committee made three trips to the Narmada Valley to meet people affected by the dam. Justice Daud submitted his report in June that year. As it says:

> The methodology adopted by us was to get the deponents ... to speak in the presence of everyone. ... [W]e see no reason to disbelieve [their depositions] particularly as the narrations were made in the presence of the Government representatives. (Daud: 4)

As you will see, it was necessary for the Committee to make that last assertion. Here are some excerpts from the report; particularly, from the depositions that were recorded:

> The government has been saying that there is land available for us, but ... there was no land for all of

us and this has been the experience of most of us. Even the government officials who accompanied us on those occasions accepted this situation ... [Even] the lands that were allotted in most of the cases were useless for cultivation and therefore many of us have returned back here.

The government had repeatedly sumitted false affidavits in the Supreme Court and all this led to the raising of the height of the dam firstly from 80 metres to 88 metres and then to 90 metres. ... They report that the rehabilitation of these PAFs has been completed. This is absolutely untrue *as we are still here in our villages* and in fact even those affected at 80 metres have not been resettled. [Emphasis added]

The *Talathi* of our area took from us a sum of Rs 1000 each and in all collected Rs 160,000 in 1987 and promised to give us *Khata Pustikas* (legal land deeds). [They] were given to us but they do not contain any entries and so they are useless pieces of paper. (Daud: 18-22)

Those three cases were recorded in villages facing submergence.

The officials have allotted uncultivable lands to many of us ... They are filled with stones and naalas and no crops can be grown on them.

Many of us are faced with the situation where the *same piece of land has been allotted to many people giving rise to conflicts*. [Emphasis added]

We also suffer from want of adequate drinking water. Sometimes the women have to fetch drinking water from a distance as far [as] 3 kms. (Daud: 23-24)

Those three were recorded in resettlement villages in Maharashtra.

The previous owners of our lands have borrowed money from the bank, so when we go to the bank for loans they ask us first to repay these amounts which we have not taken. The government has allotted these lands so these should be free of encumbrances. We have also repeatedly complained about this . . . to no effect.

We have asked the government officials time and again to give us irrigation facilities according to the rules, but that has not been done yet.

The SSP has been advertising that the new colonies would be an ideal state of life for the oustees, but if you ask us . . . *we would prefer death* [to] this kind of an unwelcome way of life. Where could we go to express our grief and grievance? Our experience is that it is of no use. Better, therefore, it is that we suffer in silence since no one has any remedy for our ills. [Emphasis added] (Daud: 25-26)

And those last three were recorded in resettlement villages in Gujarat.

Statements like these fill ten pages of the report, and even these, it says, are only 'a summary of the depositions recorded.' As Daud also writes, 'the transcripts are a repository of their hurt feelings.'

I read this record of 'hurt feelings' and wondered about the wrongs it details. But I also wondered, is it possible that all these people were lying, and that Daud and his colleagues just happened to stumble on precisely those people? That all of them have made up, or at least greatly exaggerated, their woes? For this is not an uncommon reaction among urban supporters of dams, that these complainers halt progress, object to everything.

Perhaps, but *all* of them?

Surely there is a simpler explanation: that the way Government has approached R&R has fundamental flaws.

Why must we tolerate the depressing irony that a project that is about delivering water displaces people who are then left to scrounge for that very water? Why is it not possible to actually give displaced people R&R so that they are actually better off than they were before, to the point that they will tell government-appointed committees so? Why has that never been priority number one when we build dams? Why is it instead that they are left with unheard complaints which we accuse them of making up anyway? Why must they feel 'cheated' by the government?

And if this is what governments do, and if this is a pattern that is repeated in different ways over half a century of independence, it should be no surprise that people begin to question the very idea of building dams.

•

To end this chapter, I'd like to give you a sense of what was on the minds of some very wet people in the summer of 1999.

I learned about them during a stretch of just ten minutes, if that. But it was a very unsettling ten minutes. I walked up to a second-floor flat near Churchgate in Bombay to watch a bit of video, shot just a couple of days earlier. It shook and jerked, and a fair fraction of the footage was—as amateur efforts often are—of the floor and the sky.

But those things hardly mattered. The video also showed a number of people holding hands, singing songs and shouting slogans. Later, it showed a number of these same people being arrested. Routine stuff. About the only unusual thing was that they were standing, chest deep, in water.

Perhaps you know what this was about. The little clip was shot on the banks of the Narmada, not far upstream from the Sardar Sarovar dam. Nearly unknown to much of Bombay, or even to much of the rest of the country, a drama had been unfolding in tumbledown huts on that stretch of the Narmada. In a string of villages with names like Domkhedi, Sikka, Pipalchop and Bharad that monsoon season, hundreds of men and women spent several days standing in water. The water had been rising all those days, into their homes and up around their bodies. In the video, as I said, it was at chest level. After the shooting, the water had got up to their chins.

The water was rising because of that dam, because of heavy rains upstream. The people were standing in the water because of that dam too, because they had not been adequately, or fairly, treated by its builders. They were standing in water because they wanted to tell you—you, reading this book—what that dam is doing to their homes and lives.

Writing in this vein, I know I risk losing readers. So many are indifferent, even mildly irritated, by the Narmada issue. We've had enough of that Medha Patkar woman and her Narmada Bachao Andolan, those pestilential publicity-seekers! They are trying to take us back to the Dark Ages, aren't they? Besides, they are anti-nationals who are funded by foreigners.

Because I have heard those things, I ask a few simple questions: what publicity-seekers would spend hours, that turn to days, standing in water? Watching the water rise about their bodies? What kind of publicity do they seek, did they find, if the overwhelming majority in our largest city did not even know what was happening to them just a few hundred kilometres away? Where is this foreign funding when these people wear basic cotton clothes, when they struggle to hold on to ramshackle huts only because these are their homes?

I want to ask some not-so-simple questions too: what will make us take these people seriously? To realize that they were not playing some foreign-funded game in those rising waters, but were fighting a deadly serious battle for their lives, for a say in their lives? After all, your life is not a game. Nor is mine. Why is it so hard to accept that people in the Narmada Valley are just like us?

And that's the truth of it. Standing in water, they were really trying to tell us this: that they are Indians, just like the rest of us. Citizens just like us. Grant them that much. You might just see their concerns as they see them. For example:

One: if people are to lose their homes and land to a dam, they want to be compensated for their sacrifice. Governments make promises about such compensation, no doubt. But people now look at the record of handing out this compensation since 1947. It is not a pretty record.

Two: if after all these years of independence, people still expect that they will suffer as a result of this dam, they want a say in whether the dam is built at all. After all, this is not about a couple of eccentric holdouts, but about tens, hundreds of thousands of people. Even millions, if we take all our dams since independence,

Three: if millions are going to become destitute, these protestors want us to think about what we mean by development. None of them, let's be sure, want us to return to being cavemen. But nor do they want the form of development we have had, any more. So far, it has meant dams on their lands, submergence of their homes, ignoring their needs, all so that you and I can have electricity and water in our far-off city flats. Enough of this kind of development, those wet people were saying.

Four: if they must 'sacrifice for the good of the nation', these people refuse to be told any more what's good for them. They want to be able to decide for themselves. They are tired of being those sacrificing souls. What's more, they know now that the 'good of the nation' must mean their good too. Or it means nothing.

These are some reasons people in those villages protested by standing in the Narmada's rising waters, that 1999 monsoon. They are not there any more, but even so, you might give them a thought.

●

The International Labour Organization's (ILOs) Convention 107 is concerned with 'the Protection and Integration of Indigenous and other Tribal and Semi-Tribal Populations in Independent Countries'. India was one of the earliest countries to ratify Convention 107, on 29 September 1958. Article 12 of that Convention addresses the removal of tribal populations 'in the interest of national economic development.' Such interest, we hear said so often, is the reason for dams like Sardar Sarovar. Such removed people, the Convention requires of signatory nations, 'shall be provided with lands of quality at least equal to that of the lands previously occupied by them, suitable to provide for their present needs and future development'.

The several hundred villagers who made way for Kevadia, the people who stayed put as the Narmada rose to their chins, and many others who have been touched by dams on the Narmada, they may not know about that Convention. But you, aware of what has happened to them, can deduce just how much our signature on that ILO document means.

One more reason to ask why we continue to build dams.

6

Damned by Dams

In the summer of 2001, the massive floods in Orissa showed up the keen dilemma dams can pose. By the time this book sees print, those floods will no doubt be just another incident in the endless cycle of natural calamities that have struck many parts of India, and Orissa in particular. Before the floods, late in 2000, vast areas of Orissa reeled under a major drought; before *that*, people in the state had to put their lives together after two terrifying cyclones in October 1999.

And in 2001, they had floods.

Orissa is home to the Hirakud dam on the Mahanadi river, independent India's first large dam project. In 1946, when Sir Hawthorne Lewis, the governor of the province, was laying the foundation stone of Hirakud, he expressed the fond hope that it would 'banish' from Orissa the pestilences of 'flood, drought and famine'. Today, as Orissa stumbles from cyclone to drought to flood to whatever's next, we know just how successful the dam has been in fulfilling Lewis's hope.

What happened during the floods of 2001 was more

than just a river flooding. As Rohan D'Souza wrote in the *Telegraph* (25 July 2001):

> The Hirakud dam, in fact, was one of independent India's first multipurpose river valley schemes with *flood control* as its primary objective, and which in the initial period of its construction was touted as a 'permanent solution' towards preventing floods. [Emphasis added]

Given that history, it is ironic to read about the role the Hirakud dam played in the floods of 2001. Torrential rain in the upstream catchment areas of the Mahanadi river in early July took the water level in the Hirakud reservoir alarmingly close to the full reservoir level (FRL) of 630 feet. Even with fifty-six of its sixty-four gates open on July 19, more water was flowing into the reservoir than out through the gates. The new moon on July 20 caused especially high tides along the coast. This further compounded the difficulty of releasing water through the gates. For how, and to where, would that water flow? State authorities became seriously concerned for the dam itself, worried whether it would hold up—and remember that Hirakud is an earthen dam—against this enormous pressure of water. Orissa's chief secretary, D.P. Bagchi, told the press that 'the dam's safety [is] of prime importance.' The consequences of its giving way would be appalling.

And so the gates had to remain open. The water flowing through the dam severely exacerbated an already desperate flood situation downstream, in the river's delta. 'In effect', wrote D'Souza, 'the Hirakud dam, instead of holding back flood waters, is now copiously inundating the delta.'

It's not a dilemma you'd wish anyone to face, least of all during a crisis. Open the gates, you compound the floods. Do nothing, the dam is endangered. And all the while, the rain continues, worsening the problem by the minute. And it's a time of particularly high tides. To think all this happened with a dam that was supposed to *control* floods!

Or so it would seem, so it was sold to us. Again, D'Souza writes:

> In Orissa, the prescient flood committee of 1928 noted early on that floods were inevitable in a deltaic country and it was 'useless' to attempt to thwart the 'workings of nature' through flood control measures. The committee further argued that in Orissa the problem was not how to prevent floods but how to pass them as quickly as possible to the sea and therefore, the solution lay in 'removing all obstacles' from the path of the flood waters. The report of the 1928 committee, however, was buried by the politics of the period, which then instead facilitated the construction of the Hirakud dam.

Politics, again: another example of the intricate link between dams and politics.

This is not to say this is true of every dam built in India. But the lesson from Hirakud is that we must retain some degree of scepticism about the claims made for such dams.

After all, even at its inception, there was widespread opposition to Hirakud. In 1946, 30,000 people—oustees as well as local politicians—held a demonstration against the imminent project and had to be lathi-charged. In the years that followed, there were struggles between oustees and people working on the dam. Those protests are largely forgotten today. But as Orissa faced the floods of the 2001 monsoon, we might have stopped to think: perhaps we should have listened to those protestors a little more than we did in 1946.

•

With the benefit of years living with dams, we now know a lot more about them than when they were being built fast and furious across the world. We are much more aware of the problems they bring, or compound. Other books will offer more scholarly examinations of such problems than this one. Still, in this chapter, I want to briefly examine some of them.

There has always been a lot of talk about the environmental damage the Narmada dams will cause—to the curious extent that those who criticize them are automatically called 'environmentalists' even if they discuss other issues. But there is increasing evidence that the very *benefits* dams bring themselves cause problems.

Consider some dam history.

Egypt is a largely desert country. Most of its population lives in a narrow strip of land along the Nile and its delta on the Mediterranean coast. Most of Egypt's food also comes from this narrow strip of land, about the only fertile and cultivable land in the country.

The Nile, perhaps the world's most reliable river, sustained civilization in Egypt for thousands of years. It did this by flooding Egypt's arable land every spring. The floods would carry off salts deposited the previous year and bring a fresh layer of silt. Farmers planted crops that thrived on this yearly supply of perfect soil. In the Mediterranean, a huge sardine fishery flourished, nourished every year by the Nile's flooding. Egypt was able to feed a large and growing population from its relatively small area of fertile land.

In the 1960s, all this suddenly changed. Gamal Abdel

Nasser, caught up, as Jawaharlal Nehru was, in the grand vision of technology as the nation's destiny, decided to dam the Nile at Aswan.

The result is what Marc Reisner calls 'the worst ecological mistake perpetrated in one place by mankind.' (Reisner: 487) The reliable spring floods are now a thing of the past. Bilharzia—or schistosomiasis—a disease that comes from snails in stagnant water, is rampant. The Mediterranean sardine fishery is near extinction. The rich silt no longer comes every year. In fact, the reservoir at Aswan is silting up rapidly. That's right: instead of being deposited in arable land, where it used to be so beneficial, the silt collects up against the dam. With water available abundantly and cheaply, farmers have taken to reckless irrigation. Water tables are thus turning salty and rising as well, waterlogging the soil in the irrigated areas, making them unfit for productive cultivation that Egypt once knew.

Egypt is now forced to put drainage systems in place to remove the water from irrigation canals that is seeping into the water tables. This is hard to bear, at least partly because bilharzia is now a national epidemic that causes a drain of hundreds of millions of dollars every year. In fact, Reisner believes that the bill for all these misfortunes will easily eclipse the value of the irrigation 'miracle' wrought by Aswan.

It's easy to dismiss this disastrous tale as being unique to Egypt. The truth is, it isn't. Siltation is the inexorable enemy of dams, the one phenomenon all dams are vulnerable to. Dam reservoirs throughout the world are silting up, some faster than others. And given the tremendous deforestation and consequent erosion that is taking place in upriver areas, especially in poorer countries, the problem of silting is growing more acute still. (Every picture of the flood waters that rampage through Orissa, or nearly anywhere, show that they are brown with soil from the erosion of deforested areas in the hills). The useful life of dams is regularly being revised downwards because their reservoirs silt up so fast. India's Tehri dam had its projected useful life reduced from a hundred to thirty years because of siltation (Reisner: 490). The estimated life of Hirakud itself was revised from 110 to thirty-five years, of Bhakra, from eighty-eight to forty-seven years. The Sanmenxia Reservoir in China, an extreme case, was completed in 1960; three years later, so much silt had piled up against the dam that the river threatened cities upstream. (McCully: 120)

How do you handle the siltation problem? Some people suggest a series of smaller silt-retention dams to trap the silt at various points upriver from the main dam. Even assuming that building them makes economic sense, these smaller reservoirs will, of course, silt up fairly quickly too. Los Angeles has some experience at desilting reservoirs, but the cost has been prohibitive. And there is the problem of finding a place to put all that mud: unless it is transported and thrown into the sea—at a frightful cost— it eventually washes right back into the rivers anyway.

Irrigation, far from being an unvarnished boon, has its own attendant problems. To understand why, we need to realize how profoundly unnatural irrigation is. Once water is used for irrigation, it has to find an outlet. If it drains quickly back into the river, that's fine. But if it does not drain, or drains slowly, you have problems. It percolates through saline or alkaline soils, picking up salts, and then returns to the river. The water is diverted downstream, used again, and returned to the river. On the Colorado river in the US, for example, the same water may be used *eighteen* times over for irrigation. (Which is itself an indication of how heavily exploited the Colorado is). The water also sits in reservoirs for long spells, where large amounts are lost to evaporation. The cycle continues: salts are picked up from the soil, evaporation takes away fresh water, more salts are picked up, and more fresh water evaporates. In the absence of dams and reservoirs, these salts would simply be flushed out to sea every year (which, incidentally, is why sea water is salty)—as the Nile used to do in Egypt.

But the salinity level of irrigation water keeps rising, spelling danger to fields under irrigation. Near the source of the Colorado river, there is only a trace of salt in the water. But 120 miles downstream, after the Colorado has passed through heavily irrigated land, the salinity level shoots up to about 2,200 parts per million (ppm). 2,200 ppm spells death to most crops. In the US, tens of thousands of acres of productive land is turning useless—permanently—for cultivation, covered with salt deposits. In Iraq, over 20 per cent of the arable land is permanently destroyed because of salination from irrigation.

Writing about this in *Scientific American*, a water engineer said: '[Egypt] is now faced with the universal problem of keeping salts from accumulating in the irrigated fields.' (Pillsbury) The Aswan dam, that mighty vision of Nasser's, now threatens an ancient civilization.

Irrigation also acts as a disincentive to conserve water. Since water is available abundantly and cheaply, as in Egypt, the farmers who benefit cannot, in fact, afford to conserve it. And the low price of the water is a political imperative—politics again—which justifies the building of the dams in the first place.

Besides, with the abundance of cheap water from dams removing any incentive for researching lower-cost alternatives, other efficient irrigation systems have remained prohibitively expensive. The Israelis have developed and installed drip-irrigation systems, and have pioneered many other water-saving innovations; all because they have so little water. Wasting it is almost a crime. But the large dams the rest of the world persists in building make so much water available that much of it, in fact, gets wasted. As you have seen before, a major reason dams are proposed is the desire to make productive use of water that is otherwise 'wasted' by flowing out to sea. Perversely, by subsidizing the cost of irrigation water to users, the water gets wasted anyway. And it causes long-term harm as well.

●

Consider these lines:

> In spite of non-compliance with [World] Bank resettlement and environmental requirements, the Sardar Sarovar Projects are proceeding—in the words of Chief Minister Patel of Gujarat—as an 'article of faith.' It seems clear that engineering and economic imperatives have driven the Projects to the exclusion of human and environmental concerns.

Is that an excerpt from a statement by the NBA? Perhaps by some radical environmental group? Maybe even by that ultimate bogeyman of the Government of India, Amnesty International?

No, the two sentences are from the Letter to the (then) President of the World Bank which forms a preface to the Morse Report. (Morse: xxiv)

Makes you wonder, doesn't it? Who really did Morse conclude was benefiting from these projects? Clearly not the people being displaced. Not the supposed beneficiaries, given the lackadaisical research that the Morse Committee uncovered into how and when they will benefit. At the time I first read the Report, the then Gujarat chief minister, Chimanbhai Patel, and others in his government were mincing no words in vowing that Sardar Sarovar would be built, even without World Bank assistance. In making those promises, I had to believe, they were vocal and sincere. Where, then, I wondered, was that sincerity when it came to studying the issues raised by the building of the dam, issues that the very agreements that they had signed bound them to address? Why was there just one thing they seemed to be doing with any alacrity: building Sardar Sarovar?

Reading the Morse Report is a revealing exercise, even if you believe dams are beneficial and benevolent.

Right at the beginning, the Report acknowledges the 'open and full participation' during the review by officials of all three state governments, and 'especially' by the SSNNL. I mention this to emphasize that when the review was under way, these officials did indeed extend extraordinary and generous cooperation to the Committee. So when the Report was published and was critical of their own work, they could hardly claim that they had not been heard. Instead of that time-honoured evasion, they had to find other ways of rebutting its findings. Which they did. 'Morse exceeded his Terms of Reference', they said. Morse is 'biased', they said.

Was this really so? And was it incidental that the dam-builders had made little progress till then on anything except building the dam?

The story is much the same, whichever issue Morse considered. Take irrigation. The Gujarat government divided the Sardar Sarovar command area—the area that the dam will deliver water to via canals—into thirteen regions, based on 'agro-climatic and socioeconomic features'. At the time of the Morse review, the Government had not conducted studies on the effect irrigation would have on these regions. But the Morse Committee did just such an assessment. They considered rainfall patterns, soil types, and other factors. Here is some of what they found:

Regions 1, 2 and 3 (24 per cent of the area to be irrigated): 'Canal irrigation is likely to result in serious soil degradation ... and a decline in agricultural production. [Ground water] offers additional potential in all three regions.'

Regions 4 and 7 (12 per cent): 'Both [regions] ... have been recognized as problem areas for irrigation. Only a small portion of the area seems irrigable. The proposal is to fill dugouts with canal water to supply farmers ... The average rainfall in either area suggests that the dug out tanks could be filled without the canal in many, if not most, circumstances.'

Regions 5 and 6 (16 per cent): '... [the] study of drainage conditions indicates that a considerable quantity of water is lost as runoff ... it could perhaps be utilized. In region 5, most of the area is already irrigated by wells; the ground water is of good quality ... Waterlogging and salinity would likely be a problem in the hot months.'

Regions 8 and 9 (13 per cent): 'Region 9 ... has only been partly surveyed and the area surveyed ... [presents] moderate limitations for sustained use under irrigation. In both Region 8 and Region 9, provision of canal water is likely to present waterlogging and salinity problems.'

Regions 10 to 13 (35 per cent): 'Much of the region has not been surveyed, though this is the area where most of the soils [are] unsuitable for irrigation ... In all four regions the proximity of the Rann and Gulf

of Kutchch presents problems for agriculture. The
vulnerability to waterlogging and the liability to
flooding during the monsoon will likely result in
degradation and lower agricultural production.'
(Morse: 311-312)

No doubt the Gujarat government would dispute these
assessments of irrigation potential. But surely that itself
argues that they must come forth with their own studies,
their own assessments? Strangely, they did not, nor did
they appear interested in doing so in response to Morse's
findings.

The Committee also studied the Tawa project, the first
major irrigation development in the Narmada basin, and
two other projects adjacent to the SSP: the Ukai on the
Tapi River and the Kadana on the Mahi River. 'In all three
projects,' the Report says, 'the damage from waterlogging
was significant and had been consistently underestimated
in design.' (Morse: 313) (Recall the wildly varying official
estimates of waterlogging in these projects that you read
about in Chapter 3). As a result, Morse observed, some of
the best agricultural lands in these areas were actually
going out of cultivation.

The Report concluded its review of the command area
by warning:

> The Sardar Sarovar Projects are likely to perpetuate
> many of the features that the Bank has documented
> as diminishing the performance of the agricultural
> sector in India in the past. (Morse: 321)

For my money, those are the most startling thirty words
written anywhere about this dam. I remember asking
myself as I read them: Can we afford this any longer?
When hunger remains a daily reality for so many Indians,
when agricultural land is already at a premium, how much

longer can we afford to let more of it go *out* of productive use? Is there any excuse for further 'diminishing the performance of the agricultural sector' in our country?

Or is this a naïve way to look at the issue? If so, why has it not been addressed more convincingly by those who build this dam?

But perhaps there is an excuse for this 'diminished' agricultural performance. Sardar Sarovar will also provide drinking water to millions in Kutch and Saurashtra. So we have heard, and many times, from Chimanbhai Patel, his successors and their colleagues in government. (Recall my examination of this aspect earlier in this book). Water for domestic consumption, they tell us, and especially water that is provided to Kutch and Saurashtra, is the principal justification for the Sardar Sarovar dam. Which is certainly a laudable goal.

How, then, can we explain this sentence from the Committee's conclusions: 'Despite the stated priority of delivery of drinking water, there were no plans available for review.' (Morse: 353). How is a dam described as a solution for drinking water problems when, after years in planning and construction, 'there were no [drinking water] plans available for review'?

The Report had some more about drinking water. It says that drinking water quality issues had not been addressed, nor had water disposal issues, nor the energy requirements. But what troubled me most of all was the acknowledgement by the chairman of the SSNNL to the Morse Committee that the number of villages to be served with drinking water in Saurashtra and Kutch 'are statistical figures which include 236 uninhabited villages'. (Morse: 319)

What is the possible justification for including 'uninhabited villages' in these figures? After all, if opponents of the dam claimed that its reservoir would inundate a certain large number of villages 'which include 236 uninhabited villages', they would be laughed out of the entire debate, and deservedly so. Why does the chairman of the SSNNL make such a claim and still expect to be taken seriously?

Morse also commented on the Gujarat government's apparent disregard of public health issues raised by the projects. Irrigation projects are known to carry health risks—these have been documented in India for decades. For example, Morse quotes a 1938 paper discussing 'Malaria Due to Defective and Untidy Irrigation', that blamed malaria around some of these projects on:

> Improper siting and housing . . . uncontrolled jungle clearing . . . obstruction of natural drainage by road, railway and canal embankments, with culverts too few and too high; impounding of water without regard to leakages, seepages and raised water-table levels; irrigation without drainage. (Morse: 323)

Things have not changed much since 1938. Malaria, filaria and schistosomiasis—as in Egypt—have already been identified as the three principal diseases that threaten public health as a result of the construction of Sardar Sarovar. And yet, the Morse Report says:

> We cannot explain the discrepancy in the health aspects of the Sardar Sarovar between what is known and required by both the Bank and India on the one hand, and what is not being done on the other. (Morse: 328)

For by the time the review was done, there had already been a sharp increase in malaria-induced fever in the areas around the dam. At the dam site and nearby villages, the incidence of malaria was already nearly double that in the other villages in the area. Canal construction had created stagnant pools where mosquitoes bred rapidly. Several people in the area had died from malaria. That the plans for Sardar Sarovar had not included preventive measures for malaria and other water-borne diseases seems utterly callous.

I can hardly dissect the Morse Report here at the length it warrants. But as you read it, as you absorb all that it found troubling about this dam, you cannot avoid the same question: who, really, is benefiting from the dam projects on the Narmada? 'The lifeline of Gujarat,' every politician in Gujarat calls Sardar Sarovar. But Morse makes you think: Whose lifeline is it really?

That's a good question to ask, and I will explore its implications later.

•

There is one more interesting facet to the Morse Report: the way it was received. Imagine you are a responsible chief minister. Imagine you believe the dam will greatly benefit your state. How would you respond to criticisms such as you read in the Report? By initiating studies, perhaps? By insisting that those building the dam perform the jobs they have neglected? By taking a personal interest in the progress of R&R? Or, if you think Morse was wrong, by presenting evidence to counter all the criticism in the Report? All of these? None? Something else altogether?

Don't know about you, but Chimanbhai Patel went instead to the throbbing heart of the matter, to what simply must be the most 'relevant' issue of them all.

In October 1992, he indignantly dismissed the Morse Report by saying Morse and his Committee had no right to tell us 'whether tribals are Hindu or not.'

He and his officials repeated that charge innumerable times. One such occasion was a July 1993 public meeting in Bombay that I attended. Sanat Mehta, then chairman of the SSNNL, said there that had Morse visited India before independence, he would have been 'thrashed unmercifully.' Practically apoplectic, the good chairman roared: 'How dare this foreigner tell us who is and who is not Hindu?'

You see, chief minister and chairman alike knew well: this was a convenient stick to beat the Morse Report with. Never mind that Morse wasn't pronouncing on who was or wasn't Hindu. Never mind everything else the Report discussed. Only one thing mattered—this easy way to confuse and distract people from the real issues in the project, issues that remain neglected even close to a decade after Morse submitted the Report. And no doubt, given the communally charged times we live in, then and now, chief minister and chairman alike knew just how effective this particular stick would be.

For the record, there is no point at which the Morse Report states that tribals are Hindus. Or that they are not. 'The history and customs of tribal peoples,' says the Report, 'has direct relevance to resettlement and rehabilitation policies'. In this context, the Committee studied their relationship to Hinduism. It found a great deal of overlap between their lifestyles and those of other Hindus. And even given this overlap, the Report observes that that these tribals 'did not repudiate Hinduism; rather, they affirmed their separateness.' That the 'practices [of tribal religions] have never existed entirely apart from Hinduism.' (Morse: 67-9)

What was the point Morse was making? An elementary one that anyone working with tribal peoples understands: rehabilitating such Indians can be successful only if you recognize and study their unique and distinct cultural heritage. Which, it should hardly surprise you, is something that is nearly never done. But chief minister and chairman alike preferred to characterize this as an attempt by foreigners to dictate to Indians about Hindus and Hinduism. This is what they chose to focus on, to the exclusion of all else in the Morse Report. (I'll have more to say in the next chapter about other Gujarat government objections to the Morse Report).

Somehow this focus, by itself, told me enough about the way this dam has progressed.

Dammed at Bargi

So as I trudged along that muddy trail that I have mentioned before, near the Bargi dam, both the Morse Report and the knowledge of what dams bring coursed through my mind. Because it was such a defining moment for me, as I wrote earlier, I will recount here, in some detail, my experiences then. I have always felt that what I heard, saw and felt there captures some of the spirit of the way this country 'develops'.

On 18 August 1996, a police party arrived at the village of Bijasen in Seoni district, Madhya Pradesh. They had come to arrest some of the hundreds of villagers who had been on *satyagraha*, there on the banks of the Narmada, since 21 July. In the course of that day, several villagers received injuries that they claimed came from lathi blows delivered by the police.

I was part of a three-man team asked to investigate this incident. We were invited there by the Bargi Bandh Visthapit evam Prabhavit Sangh (BBVPS), Jabalpur. While they have close links with the NBA, the BBVPS is a separate organization that has been fighting for the rights of people affected by the Bargi dam for several years. They organized the Bijasen protest in 1996.

We arrived in Jabalpur on 22 August. The next day, we travelled to Bijasen and met the villagers there. On 24 August, we met Sanjay Jha, the superintendent of police (SP) in Seoni, and Sanjay Bandhopadhyay, the district collector, at Lakhnadon town. At the Seoni district jail that day, we also met some of the fifteen people who were arrested on 18 August.

I should mention that Medha Patkar of the NBA joined the satyagrahis on 19 August, and was arrested on 20 August. However, the BBVPS told us explicitly that our mission was not to investigate her arrest, but only what they described as the lathi-charge of 18 August.

First, some background.

The Bargi dam (officially known as the Rani Avantibai Sagar Project) is part of the grand plan to develop the Narmada basin. Construction on this dam began in 1974 and was completed in 1990. It submerged nearly 27,000 hectares of land in Mandla, Seoni and Jabalpur districts; that land included 162 villages. Nearly 100,000 people were affected; about 93 per cent of them are from the backward classes (43 per cent Scheduled Tribes, 12 per cent Scheduled Castes, 38 per cent Other Backward Castes).

Before 1989, Madhya Pradesh government officials acquired land from inhabitants in these villages, offering between Rs 2,000 and Rs 3,000 an acre (or between Rs 5,000 and Rs 7,500 a hectare). This was the compensation given to people who lost, almost overnight, their entire source of livelihood: fishing in the river and farming on the rich agricultural land along its banks. Many have been forced to make a living as agricultural and manual labourers in the surrounding districts, some travelling as far as Narsingpur (75 km) for work. Several farmers are now cycle-rickshaw drivers in Jabalpur, living in slums. I met some of them there. They told me that apart from the cash compensation, they were offered no other rehabilitation.

Every year, the people affected by the dam organize protests asking for complete rehabilitation, at least as spelled out in the NWDT and the famous 'liberalized' state packages. At least going beyond cash doles. This annual affair is itself a commentary on how we pursue development.

In 1996, the BBVPS and the villagers launched their agitation on 21 July. They built a hut at Bijasen at a reservoir level of 418 metres, some 30 or 40 metres from the village itself. The location was chosen to underline their demand that until rehabilitation was properly planned and implemented, the water level in the Bargi reservoir should not be allowed to exceed 418 metres. This would allow the villagers to till the land above that level. They wanted to make the point that the loss of revenue from the reduced generation of electricity (as a result of the reservoir not being filled to its capacity at 422.76 metres) was an unavoidable consequence of the Government of Madhya Pradesh's failure to rehabilitate them. This loss could have been avoided if the government had been sincere about rehabilitation.

Because the water level stayed well below the 418 metre mark, the satyagraha remained uneventful until the middle of August. But after that, with heavy rainfall in the catchment areas of the Narmada, the water level began to rise sharply. The SP told us he was concerned that a heavy downpour might set off a sudden flood which would raise the water level dangerously. Fearing this, the collector and the SP visited the site on 17 August and sat with the satyagrahis for two or three hours, in a last-ditch effort to convince them to stop their agitation. They were unsuccessful.

•

At about 9 a.m. on 18 August, the collector and the SP returned, this time with a police force of more than 100 constables (including seven or eight lady constables) and ten or fifteen police officers. The size of the force indicated that they were intent on a showdown with the satyagrahis.

The police cordoned off the village and the SP declared the satyagraha an unlawful assembly. Fifteen satyagrahis, including three women, were arrested. As they tried to take the fifteen away, several women villagers held on to their arms. Then the police resorted to force to drive away the satyagrahis.

Pandemonium broke out in front of the satyagrahis' hut. The 500 satyagrahis assembled, including eighty or ninety women, ran about to escape the police. They allege that the police resorted to a lathi-charge. Several villagers reported seeing lathi-wielding police charging towards a group of women satyagrahis. Some of those women received lathi blows on their backs, waists, buttocks, stomachs and chests. A few were hit with rifle butts on their elbows and forearms. People related graphically how the police pushed lathis between their legs, lifted them and threw them some distance away.

Jiggelal told us that the police rushed into the hut and hit him on his arms and legs so brutally that he blacked out. Two women, Tulsibai and Manglubai, received lathi blows on their stomachs, buttocks and wrists. Rajkumaribai, one of those arrested, had an ugly lathi wound on the upper part of her thigh. Manoj Kumar Thakur, a twelve-year-old, had a wound on the left side of his face.

Then there was Jamnabai, a grandmother over seventy years old, about four feet six inches tall, bent over from arthritis, who uses a stick while walking. She showed us a large elbow wound bandaged with a leaf and some cloth. I wondered: what could cause a policeman to bring a lathi down on the arm of a tiny old woman?

•

You may dismiss this as typical of complaints from protestors who have been subjected to arrests and police force. But what did the authorities have to say?

Both the SP and the collector categorically denied having ordered a lathi-charge. There was heavy stone throwing from the villagers, the SP claimed. He explained that, as a psychological tactic to disperse the mob, the police began beating the ground with their lathis. He claimed to have restrained his men from doing anything more. But when they heard the thumping of the lathis, people started running from the hut into the village.

Then what accounted for the dozens of injuries we saw for ourselves? Both affirmed that the fleeing villagers fell on the stony ground and received various injuries. What about Manoj Kumar Thakur's face injury? That happened when he 'scraped his face on the ground.' The SP told us that it was definitely true that 'two or three' of his men must have hit people with lathis, causing injuries. But there was no need to make a big issue of this because, he said, several women in the crowd were wielding lathis against the police. He added that he himself had received, and I quote, 'a lathi blow on the thigh from a lady villager.'

The collector later corroborated this version of events in every detail but one: the blow the SP received, said the collector, was on his shoulder. Not on his thigh.

The SP's men dismantled the Bijasen hut, but only 'after the people had fled.' The collector said the hut was an 'encroachment' on government property, as all land below the full reservoir level of 422.76 metres had been acquired by the Madhya Pradesh government.

The collector was convinced there was 'mischief in the minds' of the demonstrators. They had *deliberately* chosen the inaccessible village of Bijasen for their satyagraha, he said. By doing so, they knew they would be able to 'harass Government machinery'. They 'enjoyed this thing' of harassing the Government machinery. 'Everybody', meaning the authorities, was 'running to Bijasen' and this had brought the functioning of the government in the district to a complete halt. It was 'so inconvenient' for officials and the police to make their way to Bijasen to discuss and later arrest the 'rioters', he said. (Note his use of the word 'rioters'). Why had they not staged their agitation at Bargi Nagar, where they could conveniently be met and later arrested? Still more 'mischievous', according to the collector, was that the satyagrahis made things that much more difficult for the police by pushing the women to the front when the time came to arrest them.

The SP did not complain similarly about the agitation. He admitted frankly that the villagers had 'legitimate' and 'genuine' grievances—quite an admission for a state official—but pointed out that he could not do anything to address them. His role was purely to maintain law and order, he said: his prime interest was to ensure that the people at Bijasen did not drown. In all his discussions with the protestors, this was the one point he kept making. All their other demands could be negotiated, he said: he simply wanted to be sure they would not drown.

But while saying that the demands could be negotiated, the SP added that there was simply no land available to rehabilitate these dam-affected people. Whether this was his opinion or a statement in his official capacity was unclear. The collector explained that, for various technical reasons he 'did not understand', it was not possible to restrict the water level to 418 metres when the rains began in earnest. Besides, the financial loss from reduced power generation would amount to, he said, 'a billion rupees'. This, the Madhya Pradesh government could not afford.

With this in mind, the collector said, he offered the satyagrahis a compromise package, to get them to call off the agitation. It contained two components: 1) They would be given wage employment in construction work, similar to Maharashtra's Employment Guarantee Scheme; and 2) In the December rabi season, when the reservoir level fell, farmers would be allowed to till the land below the 418 metre level.

The satyagrahis rejected this package, the collector told us.

Both officials said that they had been expecting our arrival. Since the agitation involved Medha Patkar—which, strictly, it did not, since she arrived in Bijasen *after* the 18 August incident—they knew that as soon as they made any move, the press and human rights groups would descend to investigate what had happened. Aware of this, both officials said they were 'particularly careful' in their actions.

In fact, said the SP: 'We are never as careful in other such actions as we were in this one, precisely because Medha Patkar is involved.'

Which statement really speaks for itself.

●

Our primary aim was to investigate the happenings of 18 August at Bijasen. When we got there, we saw for ourselves that too many people had too many injuries for us to accept that no force had been used that day. It is possible that a few people did fall and hurt themselves; the ground was indeed rocky. But can this explain Jamnabai's elbow wound? Manoj Kumar Thakur's face wound? The black marks and swellings so many people had on various parts of their bodies?

But the lathi-charge left unanswered the far more important question of the demands of the Bijasen satyagrahis. While a detailed analysis of all the Bargi R&R measures is beyond the scope of this book, it was hard to deny that the government had left the displaced men and women of Bijasen very dissatisfied indeed.

The irony was striking. These villagers had lost homes, land and livelihood to the waters of the Bargi reservoir; but the government, through the collector, was unwilling to seriously consider their demands. Why? Because of the consequent losses in power generation. The losses, that is, they *claimed* would happen.

In this context, consider a few quotes. These are from *A Plan For Roof*, a report on R&R at Bargi. It was written by K.C. Dubey, an ex-commissioner of Jabalpur division, and submitted to the Madhya Pradesh government in February 1987. He writes:

> The [Bargi] dam is more or less complete . . . but the plan for resettlement is being thought of now; this [amounts to putting] the cart before the horse. (Dubey: 221)

Needless to say, it was gratifying, but also deeply dismaying, to read such an admission from a Government official. Here's some more:

> The project authorities do not seem to have undertaken a comprehensive survey of the affected villages to ascertain the sectionwise and districtwise number of affected households and population.

> [The Collectors of Jabalpur, Seoni and Mandla districts] were as intrigued as I was as to why funds for resettlement of the oustees were not flowing as smoothly as [they were] in the case with the construction of the Project. (Dubey: iii)

Which reminds me of a question I have asked in this book: why has the building of the dam always been treated with urgency, while everything else in the project—R&R included—induces a certain lackadaisical approach?

But back to Mr Dubey:

> [A] project of Rs 409-412 crores has spent only Rs 16.98 lakhs (0.04 percent) on the rehabilitation of the persons who have been uprooted. And this too has been done on an ad hoc basis without any plan.

> Resettlement of the oustees is a human problem of great sensitivity and requires very careful planning and execution. The keynote of the approach to the problem has necessarily to be sympathy, understanding and complete identification with the ethos of the people who are getting displaced. (Dubey: 223 & 45)

Note Dubey's use of the term 'ethos', echoing Morse's description of the need to understand 'the history and customs of tribal peoples' to be displaced by Sardar Sarovar. Like Morse, Dubey acknowledges that when you attempt to resettle and rehabilitate people, you have to understand their ethos, their culture, their unique situation. Yet nobody has yet accused K.C. Dubey of pronouncing 'whether tribals are Hindu or not.' Far easier to sling mud like that at the foreigner.

But back to Mr Dubey again:

> The entire thrust of rehabilitation efforts should be to ensure that the people (the oustees) are as happy,

if not happier, after resettlement as they were earlier, before their lands and houses were acquired. This alone would speak of the success of the Resettlement Plan. (Dubey: 227)

Taking just that last excerpt, my experience at Bijasen only persuaded me that the Bargi dam R&R plan, if ever there was one, is a colossal failure.

•

Returning from Bijasen to Bargi Nagar, the darkness was thick and wet around us, lit only by occasional splashes of distant lightning. I felt contrary emotions: the lightning inspired awe, but there was the persistent fear of a *really* heavy downpour. The fresh air was bracing, but with every step, I wondered what I would step on next. A rock? Pile of cow dung? Crops? More mud? Perhaps a fence of thorns at the edge of the field?

In our case, we bumped into or stumbled over all of those and some other, less identifiable, objects too.

We did make it to Bargi Nagar, exhausted and caked with mud. Drinking the day's 127th cup of tea in a small *dhaba* that was lit up with several too-bright fluorescent lights, I reflected on the darkness I had just left behind. On the injured grandmother we had met in Bijasen. On that soft cough almost at my shoulder that I have mentioned before, the dog that began barking not more than a couple of feet from my ankles.

Looking back on that trip these years later, my dominant memories are of rain, mud and exhaustion. To get to Bijasen, we had tramped for several hours up hillsides, through streams and fields and bushes, down steep, muddy slopes in driving rain—all to reach a village where there was a tiny motor boat. We piled into it for the two-hour ride over the placid waters of the Narmada to

Bijasen. I was grateful for the chance to dry off in the weak early afternoon sun. But we got soaked again when returning late that night, as a storm whipped the river into an angry frenzy. And then there were the muddy hours till the cup of tea at Bargi Nagar.

To think that for the people we met in Bijasen, such an effort was an everyday affair!

The people at Bijasen were just a few of the thousands of men and women who sacrificed—that word again—so that the Bargi dam could be built. These are Indians who have never had electricity and still do not, years after the dam was completed, and half a century after independence. But they know precisely what their sacrifice has produced.

And while we could see the Bargi Nagar temple, illuminated by tubelights, for miles around as we tramped, Bijasen's own temple is visible only in the day. And only because its *shikhara* sticks out above the water. Some 100 metres from the shore. Like a lot else there, the temple drowned in the reservoir too.

All that I had learned in Bijasen, the story of lathi-charges and measly compensation and indifferent authorities—all those things might form just another paragraph in just another ordinary, unremarkable story. If what happens to the people our dams displace is tragic, sadder still is that it matters so little.

For in the villages on the banks of that stretch of the Narmada, thousands of Indians sit in the same darkness their ancestors have known for hundreds of years: the same darkness that makes a soft cough the only sign of a home, of a human presence.

7

The State Counters Its Critics

So far in this book, you've read much that criticizes the Sardar Sarovar dam, and dams in general. Such criticism is often airily dismissed as the fulminations of people with a 'bias', or of 'environmentalists', or people who 'want to hinder India's progress'—you get the idea. There is an implicit assumption in all this: that the builders of dams know just how India must progress, and are, in fact, working to bring us that progress.

I want to question that very assumption in this book.

One way to accomplish that, I've found over the years, is to read through the reactions dam-builders have to criticism. Nearly always, those reactions speak louder than the criticism itself. That is, the reactions say things about the way we have pursued development; and often, they make the points better than any criticism would.

In this chapter, I will examine in detail some reactions of dam-builders to criticism.

•

Let's start with what happened when the Morse Report came out. In an earlier chapter, I referred to the loudest public reaction to it: from the chief minister of Gujarat and the chairman of the SSNNL. Both spoke angrily about Morse's supposed propensity to pronounce on who is or is not a Hindu. But they were hardly the only people to raise this bogey.

I have with me a Government of Gujarat publication called 'Comment on the Report of the Independent Review Mission on Sardar Sarovar Project'. That is, this is the official Government of Gujarat response to Morse (I'll refer to it in this chapter as the GoGR, and to the Morse Report as the IR).

Curiously, the GoGR is unsigned. It does say in a footnote that it is 'mainly based on the report of the High Level Group headed by Prof Yoginder K. Alagh of Sardar Patel Institute of Social and Economic Research, Ahmedabad', but nowhere does it mention an author.

Anyway, the first chapter of the GoGR sings the tune the chief minister and chairman did: it bemoans the IR's 'analysis of tribals and Hindus'. This analysis is based, we read, 'on poor evidence, deliberate misrepresentation . . . and unfortunate comments [and] on all the prejudices of the colonial ethnographer's and the British administrator's perception of tribal society.' (GoGR: 5)

I have no knowledge of ethnography, so I will not comment on this any further. But I think it says something that the *first* thing the authors of the GoGR want you to read is their criticism of the IR for its comments on Hinduism. Like the then chief minister, they know how well such bogeys travel today. They figure it is far better to begin with this and get their readers' bile up, than with the much blander technical issues they might have objections to. It also means those other issues can be examined half-heartedly, which is just what reading the rest of the GoGR suggests.

•

In the GoGR, we run across this:

> Jawaharlal Nehru . . . is the butt of tasteless sarcasm
> in the first sentence of the Morse 'Independent
> Review'. (GoGR: 6)

I didn't recall this, but then I had read the IR several years
ago. Had I forgotten the sarcasm? I dug up my copy of the
IR and turned to the first sentence. In fact, here are the
first two sentences:

> After Independence, Prime Minister Jawaharlal Nehru
> ordained high dams as India's 'secular temples.' India,
> in seeking to realize Nehru's vision, has become the
> world's greatest dam builder. (IR: 3)

Try as I did, I couldn't find anything in the least sarcastic,
or tasteless, or both, in that sentence. Can you? If anything,
there's a tone of respect ('seeking to realize Nehru's
vision') here. What are the authors of the GoGR getting
at? Getting the readers' bile up again? And this time, by
telling them that Morse ridicules one of our most revered
national figures.

Further, the GoGR says:

> The Independent Review states 'In approaching our
> task, we have taken a position that our assessment is
> not simply a question of determining compliance or
> noncompliance with the Tribunal Award and loan
> agreements. The conditions laid down by the Bank
> relating to resettlement and rehabilitation and
> amelioration of environmental impact are conditions
> intended to address underlying issues.' (GoGR: 11)

So far, so good. Morse is describing how he and his
colleagues approached the conduct of their Review. But
the authors of the GoGR don't like this. The very next
paragraph, in full, is:

In view of this somewhat cavalier attitude taken by the Independent Review to its well-defined task, it becomes very difficult to discuss its findings and recommendations particularly on R&R issues. (GoGR: 11)

What cavalier attitude? What is the problem with those sentences in the IR? Or is this what I think it is, a mere cop-out? That is, they do indeed find it difficult to discuss the IR's findings on R&R—given that they are detailed and critical—and have chosen instead to blame the IR for being 'cavalier.' In any case, even if Morse and colleagues were cavalier, should the issues not be discussed?

We learn from the GoGR that

The more important aspect slurred over by the Independent Review is that an accepted R&R plan for all project affected persons accepted by all States already exists. (GoGR: 13)

'Slurred over'? The IR tells me:

[I]n the 1985 Bank credit and loan agreements relating to the Sardar Sarovar Projects, India and the three states each agreed to adopt and implement . . . R&R plans for the Sardar Sarovar dam and reservoir oustees. (IR: 28)

Later, it says:

In April 1984, the Narmada Control Authority produced a report entitled Sardar Sarovar Project: Land Acquisition and Rehabilitation of Oustees. This contains an outline rehabilitation plan . . . [etc] (IR: 46)

Still later, it says:

In 1987-88 Gujarat developed a policy for its Sardar Sarovar oustees that has since been welcomed as among the most progressive packages of measures ever devised for securing the long-term rehabilitation of people displaced by large-scale development projects. (IR: 82)

In what sense has the IR 'slurred over' the existence of R&R plans? Not only does it mention these plans, it even praises Gujarat's plan as 'among the most progressive packages ever devised' for R&R.

Yet even a cursory reading of the IR's section on R&R makes the point that the *implementation* of these plans has been shoddy and half-hearted. After all, the mere existence of a plan for R&R, slurred over or not, is no indication that it is any good, or that it will be put into practice. What the IR was looking for—what oustees and the rest of us are looking for—is evidence that R&R is being *carried out* with sympathy, efficiency and determination. That evidence was difficult to find.

•

The GoGR has an interesting discussion of the oustees from Kevadia, and the IR's examination of what happened to them. On pages 14 and 15, we find a defence of the Rs 100-250 per acre that was paid to them in 1962-63. (Was that a fair amount to pay, even in 1962-63?) It was 'consistent with similar cases pertaining to this period', we're told. Then there's this:

> The value of Rs 200 [in 1962] together with cumulative interest as applicable to Government Securities would be Rs 6,400 in 1992. (GoGR: 15)

Now the GoGR says Kevadia oustees were getting 'Rs 7,000 per acre' in 1992. Given that figure, it is hard to escape thinking that after the IR was published, some Gujarat government official was given the task of finding an equation between the 1962 payment (Rs 200) and the one in 1992 (Rs 7,000). The point, of course, was to show that Rs 200 was a fair payment in 1962. And this claim is what that official produced.

Fair enough? Except that the official, and the authors of the GoGR, clearly did not expect anyone to actually check their figures.

For 'the value of Rs 200' in 1962 to 'be Rs 6,400' (thus multiplied by a factor of 32) thirty years later, the amount of Rs 200 would have had to be invested somewhere that offers an annual interest of 12.25 per cent. The only government security I know that has consistently offered rates close to that high is the Public Provident Fund. It offered 12 per cent, though that was reduced to 11 per cent in 2000 and to 9 per cent as I write this. The problem: it was instituted only in 1968.

This is not to say that there are no government securities that might have offered 12.25 per cent over thirty years; only, I don't know of any. But my ignorance itself raises the question: which of those oustees from Kevadia could have been expected to know about these securities? If they did know, did they invest, or would they have invested in them? Is this a reasonable way to explain away the uncomfortable fact that those Kevadia villagers were merely given some cash and told to leave?

More important, is this a fair equation anyway? To compare rupee amounts from two different years, economists and others typically apply not interest rates from unnamed securities, but *inflation*: the way prices have risen. So how did prices rise between 1962 and 1992? Not 32 times, try 11. My copy of Tata's *Statistical Outline of India* tells me the wholesale price index multiplied just over 11 times between 1962 and 1992 (thus inflation over those years ran at about 8.32 per cent a year).

That is, Rs 200 in 1962 was *worth* about Rs 2,200 in

1992. And this is the money comparison the GoGR *should* have made.

Of course, we can surmise why it was not made: the GoGR claims Kevadia oustees are getting 'Rs 7,000 per acre now in addition to the payments in 1962'. Now Rs 2,200 compares somewhat unfavourably with Rs 7,000. So on seeing this Rs 7,000 in 1992, a Kevadia oustee, or maybe a Review Committee like Morse's, might legitimately ask: if you are willing to pay Rs 7,000 today, why did you offer only Rs 200 in 1962?

To which the Government of Gujarat would reply, as it has: 'The value of Rs 200 in 1962 together with cumulative interest as applicable to Government Securities would be Rs 6400 in 1992.' And Rs 6,400, of course, is close enough to Rs 7,000.

You'll recall I asked above, parenthetically: was that a fair amount to pay, even in 1962-63? My answer: no.

•

In 1981, the land in five villages around Kevadia—Panchmuli, Khalvani, Navagam, Limdi and Zer—was acquired as part of the SSP. Most of this land would be submerged by holding tanks needed for the canal; and these tanks were going to be bounded by large rock dykes. Thus these villages have come to be known as the 'rock-filled dyke villages'.

The IR discusses, at some length, the situation of these villages. Like with Kevadia, there has been much controversy about this issue. The GoGR has this complaint about the IR's treatment of the allottment of land to the people from the rock-filled dyke villages:

As on February 91 figures [sic] have been shown in Table 6.3 [of the IR]. The Executive Director of the Nigam had sent a FAX message to [the Senior Advisor

> to the Committee on resettlement] Mr [Hugh] Brody
> on March 30, '92, in which the progress of R&R up
> to January '92 was conveyed. According to this, 593
> persons were allotted land by January 1992. The
> suppression of these facts is serious since it affects the
> conclusions of the Review on the performance of the
> R&R for rockfill [sic] dyke villages. (GoGR: 22)

It is hard to know what to make of this paragraph. Given
the continuing nature of resettlement, new data will keep
coming in. This hardly precludes writing a report, as long
as it acknowledges that it is based on the data available to
date. That can hardly be described as a 'suppression' of
facts.

But apart from that, what is meant by '593 persons
were allotted land by January 1992'? A *total* of 593? That
does not change the picture Table 6.3 of the IR (see
Appendix B) itself paints, of a steadily rising number of
plots allotted to oustees. 177 were allotted by March 1989,
359 by February 1990, 466 by February 1991. Allocating
593 plots to oustees by January 1992 just continues this
trend, without being either unusually high or unusually
low. The IR itself comments on this rise just two paragraphs
earlier, when it says: 'The allottment of land to rock-filled
dyke oustees increased substantially after 1989. The figures
in Table 6.3 show the scale of increase.'

But the point the IR makes immediately after those
sentences still holds even if the Review Committee *had*
taken note of the Nigam's 593 figure. The figures, says the
Report, 'also reveal how protracted and incomplete R&R
have been for the dyke villages, at least until 1991. ...
[N]early 50 per cent of those recognized as oustees had
still not been allotted agricultural land.' (IR: 100) For
Table 6.3 also tells us that in March 1991, there were 929
recognized oustees from these villages. By that time, only

466 of them had been allotted land (the February 1991 figure above), indeed leaving 'nearly 50 per cent' without allotted land. If nearly a year later, only 593 of those 929 had received land, it seems clear that R&R had remained 'protracted and incomplete.' (Table 6.3: IR: 101)

Where is the 'suppression of facts'?

•

On environmental matters, the GoGR begins by lauding the IR's 'commendable job in listing various resolutions of regulatory and restrictive acts and rules formed by Government of India and State Governments including rules and guidelines for environmental concerns.' (GoGR: 26) It quotes the IR's complimentary remarks about various studies conducted by the M.S. University, Baroda, and the Narmada Planning Group. It quotes them because it then expresses astonishment that the IR can both compliment these studies and criticize progress on these matters:

> [A]t the onset, the Independent Review has decided 'There appears to have been an institutional numbness at the Bank and in India on environmental matters' (p 226). The logical inconsistency of the Independent Review is amazing. (GoGR: 27)

Logical inconsistency? I wanted to read about that. So I did a little digging into the IR.

Here is one IR remark, about a M.S. University study, that the GoGR quotes.

> The initial environmental evaluations done in the early 1980s, including the 1983 overview ecological studies by the M.S. University of Baroda, presented enough information for the development of proper assessment studies in the upstream area (p 274). (GoGR: 26)

You are meant to read this and ask yourself, well, then why did the IR go on to be critical? But the GoGR did not choose to quote the next sentence—yes, the *very next sentence*—in the IR. This is that sentence: '*The follow-up work has been disappointing.*' [Emphasis added] (IR: 274)

The whole thrust of the IR's comments on environmental issues is that yes, the fact all these preliminary studies were done is good, but where is the follow-up? Where is the action based on those studies? This is just why the IR comments on the 'institutional numbness at the Bank and in India to environmental matters.' (IR: 226) And you should not be surprised that the GoGR, while expressing dismay over this statement, did not choose to quote the next sentence—yes, the *very next sentence*—on that page either. This is what it says: 'The tendency seems to have been to justify rather than analyze; to react rather than anticipate.' (IR: 226)

Speaks for itself, doesn't it?

The GoGR also says that

[the Morse Committee] found that specific issues were there in 'the preliminary environmental information produced by the MS University of Baroda in 1983 after its six month benchmark study' (p 278). (GoGR: 26)

You should still not be surprised that the GoGR did not choose to quote the entire sentence from the IR that this quote is taken from. It reads like this:

Although there are a few studies or parts of studies in progress that deal with some of the specific issues, we found very little organized information apart from the preliminary environmental information produced by the MS University of Baroda in 1983 after its six month benchmark study. [Emphasis added] (IR: 278)

Logical inconsistency? Not that I can see.

Appendix B has some more examples like these from the GoGR.

•

The problem I have in dissecting the GoGR like this is that I will quickly lose my readers. That's something the authors of the GoGR knew they could count on. How much easier to quote selectively and make sly insinuations than to counter the criticism point by point! The GoGR runs to 123 pages; my examination above and in Appendix B is from just the first twenty-seven pages. If I did go on to address everything in the GoGR that must be addressed, you would have a tome in your hands, not this book. Still, I hope I have held your attention through this examination, and have also made my case: that the GoG responded to the IR—undoubtedly, a report severely critical of the SSP—in a shoddy way. Putting it kindly.

And as always, that must tell us something.

•

Then there was what happened after the Supreme Court verdict in October 2000.

Consider a sequence of events towards the end of 2000. After years of hearings, an acrimonious dispute is decided in the nation's highest court. The judgement allows the 'victors'—and I use those quotes deliberately— to resume work on a major project the case had interrupted. Naturally, they are elated. Naturally too, the 'losers' are greatly dejected. They begin exploring ways to carry on their fight, to turn reluctant public opinion their way.

Meanwhile, the 'victors' begin a concerted, if inept, campaign of abuse directed at the 'losers'.

Gathering at the project site for a celebration of the

verdict, they make snide remarks about the 'losers' and raise suspicions about their patriotism. Only, this gathering is so uncaringly and incompetently arranged that the invitees—people who are told they will actually benefit from the project—are enraged enough to beat up ministers and overturn their cars.

Then, in a large ad in a major national newspaper, the 'victors' announce that the 'losers' are passing on secrets of national importance to foreigners; that the 'losers' are supported by suspect financial transactions; and that these accusations are 'proved' by letters reproduced in the ad. Only, even a perfunctory glance at this 'proof' shows how utterly innocuous it is.

What would you think if you saw all this happening? Here's the thought that came to me: that the 'victors' know they may have 'won', but they have some noisy skeletons nevertheless, rattling away in their cupboards. So they must strike out with those trusted old weapons—innuendo, insult, lies—before the skeletons, whoever they are, come tumbling out.

This must account for the remarks made during the fiasco at the Sardar Sarovar dam site on 31 October 2000. That was when Gujarat's BJP government 'resumed' construction on the dam, following the 18 October decision of the Supreme Court. Home Minister L.K. Advani, Gujarat chief minister, Keshubhai Patel, the Narmada development minister, Jai Narain Vyas, and many others 'graced' a huge meeting at the site. Tens of thousands of less exalted Gujaratis were also there.

One Karsanbhai Patel set the tone with this observation about a childless Medha Patkar: 'She doesn't deserve any respect . . . How would Medha know the pangs of delivery?' (Quoted by Sheela Bhatt on *rediff.com*, 2 November 2000). The home minister then took the stage and wondered aloud: 'whether those opposing [such projects] were doing so . . . at the behest of some foreign nations.' After all, Mr Advani noted, these were 'the same people who had criticised the Pokhran blasts in May 1998.' (Quoted by Rathin Das in the *Hindustan Times*, 1 November 2000).

Done with their remarks, Advani, Patel and other ministers left the place by helicopter. Below, utter chaos was spreading fast. Gujarat's BJP government had used over 3,000 state transport buses to ferry some 2,50,000 people from all over the state to—'towards' is perhaps more accurate here—the dam site. Some had travelled twenty hours. But with monumental traffic problems caused by the ministers' cars parked there, many thousands were stopped nine kilometres short of the dam. They were probably the lucky ones. For those who reached the site not only had to listen to the ministers, but also had to sit through the day in the sun, with no toilet arrangements, no food and no water.

Rathin Das reports that he found some of these people 'using filthy, untranslatable and unprintable language for the BJP and its government.' They attacked two ministers and one BJP MP, overturned and set afire three official cars. As a BJP corporator from Ahmedabad told Sheela Bhatt: 'We were invited, but we were not respected. . . . Forget about facilities like toilets. Even a bottle of water was not available!' At a dam site, no less.

•

And what about the ad? Someone called V.K. Saxena, President of the National Council for Civil Liberties (NCCL) of Ahmedabad, was responsible for it. 'True face of Ms Medha Patkar and her NBA,' it proclaimed in the *Indian Express* of 10 November 2000 (p 5). It presented to credulous readers two facets of this 'true face.'

The first: 'NBA is passing on confidential documents related to projects of national importance to the [sic] foreign people.' Reading those words, you were meant to bristle with righteous rage at the NBA good-for-nothings who would hand over national secrets to foreigners. The 'proof'? An 'e-mail message' from 'Ms Chitra Rupa Palit of NBA' to two Swiss gentlemen, reproduced immediately below.

What's reproduced is handwritten, which means it could not have been an e-mail message, but let's allow that little detail to pass. The sender's name, as written on that very 'e-mail message', is 'Chittaroopa', not 'Chitra Rupa', but let's gloss over that little detail as well. Just what did Ms Palit, an NBA activist from Madhya Pradesh, write to the Swiss gentlemen? This dangerous admission: 'I am enclosing the (confidential) Risk analysis document that *we have prepared.*' [Emphasis added].

That is, Ms Palit is referring to a document the NBA has *itself* produced. How that qualifies as a national secret, Mr Saxena and the NCCL do not explain. And why should it explain? Its purpose has been served. It has painted the NBA as treasonous in your eyes. You, who saw the ad, were not meant to actually read Ms Palit's 'e-mail message.'

The second: the NBA supports itself 'through *hawala* transactions.' Hawala, as we know well, refers to undeclared transactions in foreign exchange, on foreign soil. And what's the 'proof' this time? A letter from a Nashik organization called Lok Samiti to a company, thanking the company for its donation of some thousands of rupees. (A second inference you are supposed to make is that Lok Samiti is a mere front for the NBA, one of several that allow the NBA to collect funds quietly.)

But the third line—yes, the *third* line—in that letter mentions the NBA, explicitly saying that the funds are for the NBA. In what sense, then, is the NBA taking these funds on the quiet? Besides, Lok Samiti is an Indian organization, in an Indian city, taking Indian rupees, issuing an Indian receipt, and handing the rupees over to another Indian organization. In what sense is this a transaction in another country involving undeclared foreign currency: a 'hawala transaction'? Yet why should the NCCL care about answering those questions? It has put the idea of a hawala-tainted NBA in your mind, has it not? That served its purpose yet again. No, you were not meant to actually *read* the ad.

And most telling of all is this line at the bottom of the ad: 'Space donated by a Patriot,' it says.

It makes me wonder. If patriots must seek to win public favour this way, we have slid a long way indeed from the times of Azad and Gandhi, Nehru and Patel. Someone save us from these bumbling patriots.

•

Ordinarily, this ad and the October 31 'celebrations' may not have been particularly worth noticing. After all, over the years many people have made accusations about the NBA and its activists. Besides, if a private organization makes such comments, what does that have to do with the government responding to criticism?

But just months later, there was a follow-up to these events that sort of brought everything together.

The very same NCCL submitted a memorandum to Home Minister Advani. It asked him to ban the NBA under the Unlawful Activities (Prevention) Act of 1957. An entire raft of politicians signed the memorandum to support the NCCL demand: Gujarat heavyweights Amarsinh Choudhary, Shankarsinh Vaghela, Dilip Parikh, Chhabildas Mehta and Suresh Mehta, Madhya Pradesh's Deputy Chief

Minister Jamuna Devi and the president of the Madhya Pradesh Congress Committee, Radhakishan Malviya.

As Ashish Kothari wrote in *Frontline*:

> The memorandum . . . was reportedly submitted with details of NBA's alleged subversive activities: foreign funding, passing on confidential reports related to important projects of the country to foreign agencies, human rights violations in the Narmada Valley, evasion of income tax, and letting loose a reign of violence against the project-affected persons and even government officials engaged in survey and rehabilitation work in the valley. NCCL president V.K. Saxena was also quoted as threatening to move the High Court if the Central government delayed imposing the ban on the NBA. (Kothari)

As you can see, Saxena remains confident that nobody actually read his ad and discovered how hollow are his allegations about foreign funding and passing on secrets.

I don't intend to mount a detailed defence of the NBA against this memorandum here. Still, two aspects of this episode are worth commenting on.

The first ties right in with the 'Space donated by a Patriot' line at the bottom of the ad. The whole demand for a ban implies that disagreeing with the State is not just an unlawful activity that must be prevented, it is also traitorous. Again as Kothari writes:

> There is an unwritten assumption that the state can do no wrong, and that anything it does must be in the 'national' interest. Such faith in the Indian state is indeed touching. If the interests of those behind such demands were not clear, one would even be driven to tears by such blind faith. (Kothari)

Truly: if our experience of fifty-four years tells us anything, it is that the state can indeed do wrong. You don't need to fight a dam to know that. Take a look, for example, at the roads anywhere in Bombay. Or in any city in the country. Take a look at our hospitals, our airports, our bus stations. The very authorities responsible for the state of these things speak loudest of the 'national interest' and the 'progress' of the country.

The second aspect is about the urge to lash out at those who criticize the Sardar Sarovar dam. After all, the dam-builders did win a famous 'victory' in October 2001. Why then do they feel compelled to make silly allegations and demand bans?

The only answer, it seems to me, is one I've spelled out before. The 'victors' have some interesting skeletons lying in their cupboards.

•

Nor is it just Gujarat. In an earlier chapter, I referred to a report by a group the Government of Maharashtra appointed: the Committee to Assist the Resettlement and Rehabilitation of Sardar Sarovar Project-Affected Persons, headed by retired Justice S.M. Daud. The Committee also had four Maharashtra government bureaucrats. From the Revenue & Forest department (R and FD), there was principal secretary Nand Lal, section officer K.S. Parab and joint secretary D.R. Mali. From the Irrigation department, there was chief engineer and joint secretary D.M. More.

The report is generally critical of R&R measures. Curiously, but perhaps predictably, all four bureaucrats found fault with the report and its criticisms. They submitted two notes of dissent, which are included in the report. More wrote one. Lal, Parab and Mali together wrote the other.

When I first heard about the Daud report, and particularly of these notes of dissent, I was especially interested in reading them. After all, I thought, from both my visits and other reading, I have some idea of the R&R situation. So I'm likely to agree with the major points of criticism the report makes. It would be far more interesting to read the bureaucrats' responses to the criticism.

I was not disappointed. Here are a few excerpts from the two dissenting notes (with a few more in Appendix B).

First, More's note.

Right in the beginning, More observes:

> The Committee's Report is solely based on their
> observations and submissions made to them in person
> by the affected tribals in the presence of all Committee
> Members including invitees and the field officers of
> the concerned Departments entrusted with the job of
> carrying out the R&R responsibilities. (Daud: 61)

Reading these lines, I thought to myself: so, where's the problem? This is a good description of just what the Committee should have done to assess the situation. Good, that it worked this way. Except for the presence of that word 'solely'. It makes it seem More is complaining that the Committee's efforts are superficial and sketchy—that's the tone he introduces with the use of that word. Yet what he describes is anything but sketchy. Is this to be taken as a complaint at all?

'It should be remembered,' says More, 'that the provision of irrigation along with the agricultural land is a unique feature of this R&R package which is not to be commonly seen elsewhere in the country.' (Daud: 62) Now this is hardly a criticism of Daud's report, and thus makes little sense in this note of dissent. Still, More is exactly right: and why was giving the displaced irrigated land not a priority all along? Why has it become a 'unique feature' over fifty-four years after independence, and why is it not found elsewhere? This is just the point critics have been making, isn't it? That people displaced by dams have always been treated shabbily. Here's an official admission that people displaced to build a dam do not 'commonly' get land that is itself irrigated.

The three officers from the R and FD had stronger disagreements with the Committee than More did, at least judging from the length of their report. Yet it begins on an odd note indeed: 'Shri Nand Lal ... could not participate during the visits of the Committee but still he is conversant with this subject as he has visited Rozwa on 4.2.2000.' (Daud: 65)

Nand Lal visited *one* village over a year *before* this Committee was constituted, and still he claimed to be 'conversant with this subject.' Is this how officials have treated the entire business of R&R all along? What might Nand Lal claim to know if he had actually joined Daud and his Committee on their travels?

But with that done, these officers embark on what I can only call a tyranny of 'should'. Consider:

[T]he Collector . . . has initiated process for purchasing 200 ha. of private land. We recommend that the Government *should* expedite this process . . . The Government *should* also identify other private lands. . . . [W]e recommend that percolation tanks/minor irrigation tanks as are proposed by the irrigation department *should* be completed on priority basis. . . . [W]e recommend that all the vacant posts in the Sardar Sarovar Project Affected Persons Rehabilitation Division *should* be filled urgently and very dedicated employees *should* be posted on these posts. . . . [A]ny individual grievances *should* be referred to Grievances Redressal Authority for redressal. [Emphasis added] (Daud: 66-69)

A Committee points out various problems in the implementation of government policy. The government officials' response is that government *should* do this, *should* do that, *should* do the other. But of course the government *should* do those things, the problem the Committee points out is that it is *not* doing them!

•

One last example of the State's strange reactions to criticism. Take the now-famous contempt case that was decided as I wrote this.

On 13 December 2000, a few hundred people gathered outside the Supreme Court in New Delhi. Now that the Court had ruled that construction on the Sardar Sarovar dam could resume, most of them faced the loss of their homes and lands to the dam reservoir. They had come to register their protest at the Court's decision, to tell the Court of their anguish. To tell the Court, too, that though they were being made to leave their land, they had not been given any land for R&R. This was an unequivocal failure to comply with the Court's own order. Slogans were raised that day, speeches were made. Some were critical of the Supreme Court. Others spoke of how the judges were out of touch with the situation of people in the Narmada Valley, and of people in general.

And somewhere on the sidelines, three men were drawing up a First Information Report (FIR). I can do no better than quote that FIR in its entirety:

First Information Report dated 14.12.2000

I, Jagdish Prasar, with colleagues Shri Umed Singh and Rajender were going out from Supreme Court at 7.00 pm and saw that Gate No. C was closed.

We came out from the Supreme Court premises from other path and inquired why the gate is close. The were surrounded by Prasant Bhusan, Medha Patekar and Arundhanti Roy alongwith their companion and they told Supreme Court your father's property. On this we told them they could not sit on Dharna by closing the gate. The proper place of Dharna is parliament. In the mean time Prastant Bhusan said.'You Jagdish Prasar are the tout of judiciary. Again medha said 'SALE KO JAAN SE MAAR DO (kill him). Arundhanti Roy commanded the crow that Supreme Court of India is the thief and all these are this touts. Kill them, Prasant Bhushan pulled by having caught my haired and said that if you would be seen in the Supreme Court again he would get them killed. But they were shouting inspite of the presence of S.H.O and ACP Bhaskar Tilak marg. We ran away with great with great hardship otherwise their goonda might have done some mischief because of their drunken state. Therefore, it is requested to you that proper action may be taken after registering our complaint in order to save on lives and property. We complainants will be highly obliged.

Sd.Complainants

I assure you that was verbatim. The complainants filed the FIR at the Tilak Marg police station, and much the same document was submitted to the Supreme Court as a petition, praying for action against Roy, Patkar and Bhushan.

As Roy pointed out in her reply to the Supreme Court:

The police station . . . has not registered a case [based on the FIR]. No policeman ever contacted me, there was no police investigation, no attempt to verify the charges, to find out whether the people named in the petition were present at the dharna, and whether indeed the incident described in the

FIR (on which the entire contempt petition is based) occurred at all.

And yet, the Court issued notices to Roy, Patkar and Bhushan, announcing its intention to determine if they had committed contempt by these utterances outside its gates. By itself, that would be surprising: that the Supreme Court acted on a case the police showed no interest in. But more surprising is that it actually entertained this particular petition, when even a casual reading of the FIR shows exactly how shoddy it is.

Why, when they filed their petition, were these petitioners not flung out as they deserved to be? The sole explanation that comes to mind is that the Supreme Court sought to teach the NBA supporters a lesson for their previous protests against its decision.

And that reaction from the Court fits in perfectly with the various previous ones I've discussed above. The contempt lies right there.

8

Progress as Patriotism

In this book, I have referred several times to a slim and glossy publication called *FACTS: Sardar Sarovar Project*. Written by P.A. Raj, who describes himself as a 'Part-time Technical Advisor' to the Sardar Sarovar Narmada Nigam Limited (SSNNL), *FACTS* has run into several editions over the years. I have the 1989 and 1998 versions. The book leaves no doubt that it is an attempt to put across the case for the dam, as articulated—if somewhat hazily—by its builders themselves.

A few chosen passages from the 1998 edition tell an interesting story.

The book opens with a message from Jai Narain Vyas. Vyas was a minister in Gujarat until a curious little episode in December 2000. During a cabinet meeting, Vyas told *rediff.com*, 'a heated discussion broke out between me and [chief minister Keshubhai Patel]. At one point, I told him, "You are lying!" So he asked me to resign, which I did.' (*rediff.com*, 20 December 2000)

Still, at the time of P.A. Raj's 1998 edition, Vyas was Gujarat's minister for Narmada and Major Irrigation

Projects, and it is in that capacity that he sent a message to be published in this book. One line in it ran thus:

> Some activists due to their own peculiar beliefs oppose the development Projects of the country. (*FACTS*: 3)

V.B. Buch, chairman of the SSNNL at the time, wrote a preface for the book. He says:

> The [SSP] can truely [sic] be regarded as of national importance on account of great contribution it will make to improving the quality of life of rural masses ... It is imperative for the nation to make best use of [Narmada] water. It is in our contry's [sic] interest to maximise hydro power generation which is truely [sic] renewable ad [sic] indigenous source of energy. The Narmada Project assumes a special national importance in this perspective. ...
>
> I wish to dedicate this book to all those interested in the economic progress and well being of our country. (*FACTS*: 4-5)

Elsewhere in the book, you will find these statements written by Mr Raj himself:

> [The project] is a real drought proofing project and hence 'A Project of National Importance.' [p. 10]
>
> [T]he project ... is the real lifeline benefitting [sic] the entire nation incorporating the special needs of four participating States viz. Madhya Pradesh, Maharashtra, Gujarat and Rajasthan. (*FACTS*: inside back cover)

I present all these excerpts to draw attention to the not-so-subtle equation that is being spelled out in this book in particular, though it appears in many other such contexts too. The equation is: a dam on the Narmada means progress for the nation. This is a given, and you must accept it as such.

A related equation is this one: if you have doubts about the dam, you must have 'peculiar beliefs', and of course you must also 'oppose the development' of the country. It's also possible you imbibed those doubts, as Home Minister Advani mused aloud at the site of the dam on 31 October 2000, 'at the behest of some foreign nations.'

Without raising any questions, you are supposed to swallow all this as truth. And in fact, many people do just that. Soon after that October 2000 judgement was handed down in the Supreme Court, I wrote a column about it and expressed my dismay. As I half-expected, I was deluged with lectures about the silly notions I had about the dam.

One reader wrote: 'It is not a big deal that people have to be moved around in the name of development.' Another urged me not to hinder those who were 'building a prosperous India', for didn't I know 'progress' is 'very seldom painless'? A third explained with crystal-clear logic that people to be displaced by the dam 'also need internet' and so I should 'start supporting [the] Narmada project.'

I wondered idly—if the first man, and not some amorphous 'people', had 'to be moved around in the name of development', would he consider it 'a big deal'? I also wondered, though I didn't attempt to comprehend, just how the Narmada projects would inflict the Internet on those they displace.

To many people, 'development' has clearly become a motherhood and apple-pie notion—a slice of a nation's soul, not to be questioned—and dams, of course, mean just such development. There is a very good reason for persuading people to believe this: if the thought that a dam is progress becomes part of your consciousness, the national consciousness, if as a citizen your first thought about a dam is that it symbolizes 'development', it becomes very hard to see anything wrong with that dam. And then those who object can easily be branded 'anti-progress', accused of destroying the 'well being of our country', and—why not?—called 'anti-national' as well.

Yes, by this logic, if you think you are uniquely 'interested in the economic progress and well being of our country', you will find the people who oppose the dam odious traitors indeed.

For all its varied mistakes and almost studied mediocrity, this *FACTS* publication maintains one primary thrust very well. That is, to paint the opponents of the dam as subversives who are pulling India off its glorious path to progress. At one point, Raj even makes it quite clear what he thinks of the opponents:

> The anti-project lobby is opposing the SSP on almost each and every count. They are trying to halt the project by seeking help from foreing [sic] countries. (*FACTS*: 48)

There it is, finally down in black and white: that age-old, ubiquitous, never-fail 'foreign' bogey. Believing that he has successfully made the point, Raj doesn't even bother to substantiate his accusation about 'foreign countries' here. Easier to just say it and move on.

And yet, these subtle implications and assumptions must, indeed, be questioned. Not just because we are a democracy and deserve more than to have things told to us. No, it's actually the other way around: asking questions is the foundation of democracy. And with dams and development in particular, this idea of the national progress is now, after half a century and more, shaky enough that it can, and must, be questioned. What's more, while doing so we must actively spread the idea that, far from undermining a nation, such questioning strengthens it.

What is this progress, after all? Whose is it? Whom does it benefit? What has been its record these fifty-five years? Given our record, given how we have 'developed', just what does 'development' mean anyhow? Just what is a 'project of national importance'? Who decides, and how?

These are not idle questions asked in a vacuum. The truth about India is, more and more people are starting to ask these questions. Increasingly, people are no longer satisfied with the bland homilies that pass for answers. Simply labelling something a 'Project of National Importance' means little. But if it is to mean anything at all, it is that the project concerned is important to you and to me, ordinary citizens of India.

•

The term 'development', in the sense many have come to understand it today, in the sense that some countries are 'developed' and others are not, probably has its roots in President Harry Truman's Inaugural Address in 1949. As such occasions seem to demand, Truman sought to define some grand purpose, some overarching vision, for his term in office. It couldn't have been difficult—a long and destructive war, that had left the world weary of bloodshed, still dominated public memory. Sentiments promoting reconstruction and renewal, and America's role in leading the way, must have been welcome, and found their way into his Inaugural speech. So Truman said:

> We must embark on a bold new program for making
> the benefits of our scientific advances and industrial
> progress available for the improvement and growth
> of underdeveloped areas [of the world]. (Sachs: 9)

At one stroke, Truman generously offered to share with the world all that made the US a wealthy and powerful nation; he also drew a line that separated the 'developed' and the 'underdeveloped' parts of the globe. That line, and those terms, have since sunk deep into minds all over the world. Countries think of themselves and others as 'developed' or 'developing' or 'underdeveloped'—though over the years that last term has become something of a pejorative. In much the same way, the planet was divided into the 'first' world and the 'third' world (who even remembers what happened to the 'second' world?); 'third world' now is a synonym for 'underdeveloped'.

Left unexamined to any depth was the notion that 'development' automatically means 'scientific advances and industrial progress', at least as seen in the US. Was that true? Was the US—the West in general—the example for the rest of the world to emulate?

This is not to belittle the US, nor to deny that it is an admirable place when you consider the life it is able to offer its citizens. If India, and many other countries, can aspire to give their citizens the opportunities Americans enjoy, that would be a fine thing. But even so, we do need to ask: is 'development' as it has happened in the US truly possible in other countries? Is it feasible, desirable, wise, to follow that path? Even if we do follow it, should we do so blindly? Might it be better to consider local conditions, learn from the mistakes the US has also made on the road to becoming 'developed', and avoid those pitfalls?

Assuming that we do ask all these questions, one remains that could overturn all the others. Yes, it is good to aspire to give all Indians the kind of opportunities Americans enjoy. But are those who decide these things in India truly imbued with that aspiration? Do those who speak of 'progress' and 'development' for India truly want it for all Indians?

An honest answer to that, it seems to me, would explain much of what's wrong in India today. It would also explain the cynicism about such 'progress' that greets such publications as *FACTS*, or that you probably detect running through this book. After all, we have heard platitudes about 'development' for many years, from all manner of eminent Indians. Yet we still have countless millions who thirst for water every day; we have the world's largest number of illiterates; we are content that seventy per cent of our population doesn't have access to something as ordinary as sanitation. We have pervasive corruption, often practised by the very Indians who are most vocal about 'progress.'

Wolfgang Sachs alludes to some of this cynicism:

> The idea of development stands like a ruin in the intellectual landscape. Delusion and disappointment, failures and crimes have been the steady companions of development and they tell a common story: it did not work. . . . Nevertheless, the ruin stands there and still dominates the scenery like a landmark. Though doubts are mounting and uneasiness is widely felt, development talk still pervades not only official declarations but even the language of grassroots movements. It is time to dismantle this mental structure. (Sachs: 1-2)

So when *FACTS* talks about 'progress', it is hard not to treat that with some scepticism. I would even argue that the scepticism is a good thing; had we seen more of it earlier, we might have done a few things right.

●

Let's begin asking some of those questions in the context of dams on the Narmada. First, what progress?

Despite all of India's hectic dam-building since independence, in the early twenty-first century we actually have more land in India that is considered prone to drought and flood than we had in 1947 (Roy). Despite all the years of rhetoric about how those dams would make water available to Indians who crave it, today we have some 170 million Indians—think of it, over half the total of Indians we had at independence—who cannot count on clean drinking water. (Human Development Report: 146). Only 15.3 per cent of Indian households have a toilet (Shariff: 241); one consequence is that for about 700 million Indians—70 per cent of us—basic sanitation is nonexistent (World Development Report: 224).

For the millions of Indians clubbed together in that last paragraph, therefore, access to water is a major daily worry.

And I hope you don't need a figure from me to comprehend how many Indians go hungry today. Walk out on any street, in any city in India, at any time of day, to get an idea. People are hungry even though we claim to be not just self-sufficient in food, but actually producing food surpluses every year.

All this defines our country today, and we have arrived here 'in the name of development'. Following some ordained path to 'progress'.

Caveat: This is not to say that India has not progressed at all since 1947. After all, we *are* producing those food surpluses. We may have more illiterates than we had Indians in 1947, but the level of illiteracy—the fraction of our population that is illiterate—has certainly gone down. In our cities, we can usually count on piped water supplies, often reliable electricity and working phones. We have made progress.

So let's ask a second set of questions then: whose

progress? Who has progressed? Or let's put it another way: who has not progressed?

I'll answer that first of all with some pertinent figures I dug up for an article I wrote in 1996.

> Number of people displaced by 'development' projects since independence (estimated by *India Today*): 30 million.
>
> Percentage of that total resettled: under 10.
>
> Percentage of India's tribal population displaced by development (estimated by Lokayan): 20.
>
> Number of people who will be displaced by Narmada dams: estimates range from 100,000 to 1,000,000. (D'Souza, 1996)

What do these figures tell us? First of all, that resettling people displaced by projects that supposedly further the country's 'development' has never been much of a priority. Second, that tribals figure prominently among these displaced people.

In other words, there is a price we have paid for the progress that many of the readers of this book will acknowledge their country has made. There is a price we have paid for the benefits, the comforts of life, that many of the readers of this book enjoy. And yet, even using the word 'we' in those last sentences grates. Because it is really not 'we' who have paid that price, in the sense that is really not people like you and me.

In general, it is *other* Indians—many of whom are tribals—who have paid the price. And most of them certainly do not 'enjoy' too many comforts of life.

To understand this point, consider the dam on the Narmada I have mentioned before, the Bargi that's south of Jabalpur. When the Bargi dam was built, its reservoir submerged 162 villages that lay upriver from the site. Tens of thousands of people who used to live in those villages lost their homes and had to move. When you speak to them, as we did at Bijasen and Jabalpur in 1996, you learn some quite unsavoury things.

Some of them actually moved twice: the reservoir submerged even the land the government settled them on the first time. When they lost their land to the dam, the government gave some—only some—of the villagers cash as compensation. Substantial portions of even those amounts were lost, as officials demanded bribes before handing over the cash. Any other compensation for these villagers' losses remains just an empty promise.

Today, many of the men of those 162 villages must travel hundreds of miles to fields in distant parts of the state to find farm work. Others, including several once prosperous farmers, now live in a putrid garbage-infested slum in Jabalpur, pulling cycle-rickshaws for a living. This is what the country's 'progress' has meant for these people.

I don't know of any studies that have proved this, but there is enough anecdotal evidence (like my experiences in Bargi) to suggest that large numbers of people displaced by dams do end up in city slums, on city streets. The irony that compounds their trauma is that we urban and middle-class Indians want to wish away this phenomenon, to push those slum-dwellers somewhere outside our cities, to prevent them from arriving in the first place. Not only must they 'sacrifice' in the 'national interest' to make lives like yours and mine better, they must not defile our cities either. No matter that where they come from, opportunity is nonexistent.

Yes, this is what 'progress' and 'development' have meant to millions of Indians ever since independence. No wonder a slogan that often resounds through the Narmada

Valley is this one: '*Vikas chahiye, vinash nahin.*' ('We want progress, not destruction.')

A third, but as vital, question is: who says these things about progress? Who decides that a project is of National Importance, important enough to capitalize those two words? Do they really care for the national interest?

Let me tease out this point further. Suppose Sukh Ram, once a central minister for telecommunications, was to announce that a dam was to be built in his state of Himachal Pradesh. Suppose he went on to say that this was a project of national importance that would further economic progress. Suppose he also pronounced that those who objected to this new dam were blocking economic progress and hindering the nation.

Listening to him say all this, and remembering the charges made against him in 1996, how seriously might we take his pronouncements about progress and national importance? If we cannot trust our leaders with public money, how can we trust them when they pronounce on the country's development? What development can we expect from people who have been accused of crimes?

For far too many years, our leaders have simply pronounced on progress and development, and what is or isn't the national interest. For just too long, we have shirked our responsibility to question those pronouncements. The result is the India we see around us: one in which a fits-and-starts kind of 'development' has resulted in our awful roads, seedy airports, crumbling bridges, flyovers that worsen traffic, pollution everywhere (recall the Gujarat Ecology Commission's report that described Gujarat as the 'most polluted state'), and widespread poverty that threatens every day to wipe out any progress we make.

Development, as we have known it, has left behind hundreds of millions of Indians.

It's time to demand better, and that must happen by asking questions about the people who lead us, about the 'development' they speak of. Remember what Sachs says:

> Delusion and disappointment, failures and crimes have been the steady companions of development and they tell a common story: it did not work. (Sachs: 1)

Because it did not work, many in India are profoundly disappointed with development as it has been sold to us these fifty plus years. It has brought inequality and corruption; development has certainly deluded us. It makes us wonder: Should we keep pursuing Truman's idea of development? Or is there another way?

•

At least with dams, there are other ways. Chapter 4 explored some thoughts about one, which would not deprive Gujarat of any water. But I would like to go beyond just changed parameters—even greatly changed parameters—in the finished product while searching for another possible way. I mean that the process itself should stand a review.

Particularly because we are a democracy, at least in theory, it is criminal to ignore the wishes of the tens of thousands whom the dam will affect, and go ahead with projects. I say this mindful of the equation I referred to in an earlier chapter: that if the dam affects some people, it will benefit many more. Enough of those affected people are now asking about dams and the sacrifices they made, and it is time to listen. I believe we must listen to them even if dam-builders contend that many more will benefit, not least because that claim itself is contentious and so must be challenged. The case I describe below will show

how fallible is the entire idea of costs versus benefits, and the way it is put forward nevertheless.

•

Ever since the NWDT Award was handed down, since the construction on the Sardar Sarovar dam began, since the Supreme Court decision—through all this, there has been a sort of steady muttering from Madhya Pradesh. It has generally been ignored, but it persists.

What's the muttering about? A couple of issues.

One, a fear that Madhya Pradesh simply has no land to offer people who are being displaced by the dam's reservoir. Madhya Pradesh has the greatest number of oustees, and even if Gujarat has offered to settle them there, that's no reassurance. That's because large numbers are involved, not all the displaced people will want to—or can—move to Gujarat, and, therefore, Madhya Pradesh will have to provide them with adequate R&R.

Two, knowing all that, Madhya Pradesh has consistently asked for a lowered final height of the dam: 436 feet as opposed to 455. This, claims Madhya Pradesh, will substantially reduce displacement within the state without reducing Gujarat's quantum of water from the dam. Therefore, it will result in a much lowered burden of R&R.

In 1994, Madhya Pradesh's Chief Minister Digvijay Singh wrote a letter to the then prime minister, Narasimha Rao, and made these points like this:

> [R]educing the height of the dam from 455 ft to 436 ft ... will have the following advantages: ...
>
> 2. About 38,000 people would be saved from displacement in Madhya Pradesh alone. Resettlement and rehabilitation of such large numbers is always

difficult especially when the population being displaced belong largely to the tribes who have strong community bonds.

3. About 25,000 acres of land that could be saved from submergence by such reduction includes some of the more productive land in the Nimad area of Madhya Pradesh. . . .

9. The basic case is that the Tribunal, when it made its award, the context was very different [sic]. The States underestimated both the number of people affected by submergence and the misery associated with it. Madhya Pradesh earlier estimated only 6,147 families as being affected whereas today it has gone up to 33,014 families with a population of 114,000. If this extent of submergence was known, perhaps the Tribunal would not have recommended the present height [455 ft]. (Singh)

I find this a most interesting letter on various levels. First of all, there's the very assertion that the height reduction will have 'advantages'. This has never been a popular thing to advocate, yet Digvijay Singh both says it and lists such 'advantages'.

Second, there's the concern for the people to be displaced, even the use of that word 'saved' in point number 2. R&R is 'always difficult', wrote Digvijay, which may qualify as the understatement of 1994.

Third, there's Digvijay's concern for the loss of what he calls 'productive land in the Nimad area'. An often-made argument is that far more land will be turned fertile by the dam than is being submerged. In this letter is a hint that Digvijay does not quite believe that argument, especially when applied to Nimad, where you will find some of India's most fertile soil and some of its proudest farmers.

Fourth, the admission that the dam builders 'underestimated' the extent of the problems to be overcome, and the 'misery' that would be caused. Despite the innumerable glossy booklets and rosy claims about R&R, despite the policy statements that say R&R must aim to 'better' the lives of the displaced, a Chief Minister confesses that the dam will cause 'misery'. Besides, I am startled by the extent of this 'underestimation'. How, and why, did Madhya Pradesh undercount its affected people before the NWDT by *eighty per cent*? This difference is not explained by population growth—surely the numbers in the submergence area did not grow fivefold in a matter of a decade and a half. So where were all these people while the NWDT was deliberating? Could it be that numbers were deliberately reduced to pull wool over the NWDT's eyes? And if that is so, why shouldn't we begin doubting the terms of the Award?

Fifth and most important, there's a running undercurrent to all this that goes to the heart of the whole debate about this dam: the entire notion of costs versus benefits, the equation that I have referred to several times before. After all, we are asked to assume that the benefits of the dam far outweigh the costs. Then why did the 'States underestimate ... the number of people affected by submergence' before the Tribunal? Because the costs would, therefore, be shown to be far smaller than they are? Because if there had been more accurate estimates given to the NWDT, perhaps the benefits would not have so handily outweighed the costs?

In some ways, this is the fulcrum of this book. When our governments' own officials make these points, it calls into question any easy assumptions about costs and benefits, about the equations that are made, about the call to certain people to 'sacrifice' for that elusive 'national interest'. It tells us, therefore, that we must question the way these assumptions and statements are made, these ideas propagated.

Now I am no particular fan of Digvijay Singh. I fully believe he has his own political reasons—as opposed to a sincere concern for the displaced people or that fertile land—for writing that letter, for continuing to make the same case even today. Political leaders must and do operate that way. But that only reinforces the point: the building of this dam is driven less by cost and benefit calculations, or the needs of water-scarce areas, than by *political* imperatives.

For this reason if for no other, we should be steadily sceptical of the way this dam is being built, of facile equations that are offered to us on a plate and which, it is assumed, will be believed unquestioningly.

This is why we must listen, far more than we ever have, to those affected by dams. If their steady refrain is that the dam will reduce them to misery, that—still, as always—should tell us something.

And that, with what you've read in previous chapters, forms a fine context in which, finally, to examine that Supreme Court judgement.

9

Judgement on a Valley

For a lot of people, the Supreme Court's decision to dismiss the Narmada case, allowing construction on the Sardar Sarovar dam to resume, came as a body blow. To them, it was no less jarring than the murder of a Prince of an American Camelot was, in 1963. Yes, where were you when you heard that JFK was assassinated? When the World Trade Centre came down? Such questions are something of a cliche, even forty years after John F. Kennedy was shot. Something of the weight, the import, of these events has now passed into our collective memory. Much the same is true of other momentous events of our time: Indira's assassination, the demolition of the Babri Masjid, the Pokhran nuclear tests.

For me, one of these events took place on 18 October 2000. I remember precisely where I heard the news: in the front seat of a taxi, struggling through traffic on Bombay's Pherozeshah Mehta Road. I caught sight of a journalist friend dodging the cars, going somewhere in a great hurry. Leaning out of my window, I called out to her. She turned, saw me, and replied.

I can still remember her voice as it floated above the din on that generally noisy street: 'Haven't you heard? The Supreme Court has thrown out the Narmada case!' I can still remember slumping in my front seat, the breath suddenly knocked out of me. I can still remember trying all day to shake off the feeling of desolation.

For many in the Narmada Valley, the judgement took on an intensely personal hue, because it meant the inevitable, inexorable loss of homes and land to the dam. Elsewhere too, people mourned the decision: people who did not necessarily experience tangible losses themselves, but were nevertheless filled with a sense of betrayal. A sense that a certain faith in the processes of law and justice had been overturned.

And yet, undoubtedly, many people also thought that day in October was a signal one. To them, the judgement brought great joy rather than dismay. A large constituency—particularly in Gujarat, but spread through the rest of the country too—wants that dam built. For them, the Supreme Court had delivered justice, and emphatically. After years of delay, work on the dam would now resume. It is scarcely plausible that all these people are somehow deluded in their satisfaction. After all, they have an undying faith that the dam will bring them water. But as I said earlier, I write this particularly to reach out to such believers in Sardar Sarovar; to try to persuade them that though this judgement brought victory to those who would build the dam, it should alarm them for what it is. Indeed, it should alarm us all. Never an easy task, I know, to sell victors such a message. But in this chapter in particular, and in this book as a whole, I want to make just that case.

Especially after that October 2000 judgement, there has been much talk about respect for the judiciary, and about how criticizing the judgement might amount to contempt of court. I want to begin by examining what these two notions mean.

What, after all, is 'respect' for the judiciary? Does it mean that we must accept without question everything our judges pronounce? When I speak of respect for a colleague, or a boss, or a favourite professor, I don't mean that I simply swallow whatever they say without a murmur. In fact, if someone demanded that of me, I imagine I would have a profound disrespect for them.

When I respect someone, it's because we can exchange views and ideas in a mature, reasonable way. Because he is willing to listen to and respect what I say. Because the response I get encourages me to think, to question, to articulate and to speak my mind. These are the elements of what we call respect.

And this, in particular, is what respect for the judiciary means to me. As citizens, we show and nurture respect for our courts if we are willing to follow their deliberations, scrutinize and understand their decisions. We may not agree with such decisions, but we must understand them; and having done that, we must be persuaded that the judges acted with wisdom and justice. In a democracy, this is the minimum expected of citizens, the minimum the judiciary should hope for from citizens. To me, it is a sign of deep disrespect to simply accept a judicial pronouncement just because it has been made.

Take the US Supreme Court's split 5-4 decision in December 2000. That decision, of course, handed that country's Presidential election to George W. Bush and relegated Al Gore to a footnote in history. The closeness of the election, the fact that Gore won several hundred thousand more votes than Bush across the country, the bitterness of the tangle over recounts in Florida—these factors meant that the Supreme Court's deliberations would be very closely watched indeed, its verdict guaranteed to be a contentious one. The uncomfortable fact that the 5-4 split reflected the conservative-liberal divide among the nine Supreme Court judges only served to underline the bitterness.

Of course, the Court's decision was *accepted*, in that Gore promptly conceded the election and Bush became President. But the verdict was hardly left unscrutinized. Or uncriticized. Ronald Dworkin, Professor of Law and Philosophy at New York University, and Professor of Jurisprudence at University College, London, commented on it in a major American magazine. Consider these lines from his article:

> The conservatives [in their majority verdict] stopped the democratic process in its tracks . . . first by ordering an unjustified stay of the [Florida vote recount] and then declaring, in one of the least persuasive Supreme Court opinions that I have ever read, that there was no time left for the recount to continue. . . .
>
> [T]he Court['s] decision ensured both a Bush victory and a continuing cloud of suspicion over that victory. . . . [T]he troubling question [is whether] the decision reflected not ideological division, which is inevitable, but professional self-interest. . . .
>
> [T]he legal case [the five conservative judges] offered for crucial aspects of their decisions was exceptionally weak. . . . [Justice] Scalia argued . . . that Bush would be harmed if the recounts continued because if the Court later decided that the recount was illegal, the public's knowledge of the results would case a 'cloud' over the 'legitimacy of his election.' That bizarre claim [assumes] that the public is not to be trusted. (Dworkin)

Notice how Dworkin has used, in his dissection of the judgement, such words and phrases as 'bizarre', 'exceptionally weak', 'unjustified' and 'least persuasive.' He also makes the serious assertions that the Court 'stopped the democratic process', left a 'cloud of suspicion' over Bush's victory, and acted in 'professional self-interest.' And perhaps most damaging of all, Dworkin believes the decision reflects the Court's lack of trust in the public.

Is Dworkin therefore guilty of contempt?

Dworkin writes:

> We must try ... not to compound the injury to the Court [that the judgement has caused] with reckless accusations against any of its members. But those of us who have [argued] that the Supreme Court makes America a nation of principle have a special reason for sorrow. (Dworkin)

Among others, the writer Gore Vidal is even more forthright. Profiling him for an interview in *The New Statesman* (15 October 2001), Johann Hari writes:

> [Vidal] is also unafraid to carry on drawing attention to the *illegitimacy* of President Bush. At least five members of the Supreme Court *should have been put on trial* (for installing Bush) by the Senate, which is in charge of that under the constitution. Two certainly *should have recused themselves.* Clarence Thomas's wife was working to recruit people for the Bush administration; he *should not have sat in judgement.* Antonin Scalia's son was working for the law firm that represented Bush before the Supreme Court. *That isn't done.* Without those two, the decision would have gone for Gore. [Emphases added] (Hari)

What would happen in India to someone who suggested that the resumption of construction on the Sardar Sarovar dam is illegitimate, that the judges of the Supreme Court who allowed it should be put on trial?

To reiterate the point: it seems to me that Dworkin's and Vidal's criticisms, and many other such examinations of the US election decision—critical or otherwise—truly reflect respect for the judiciary. For they speak of a deep desire to study and understand judicial decisions, and, in fact, the judiciary itself. The judiciary, and by extension the nation, can only benefit from such a process.

Likewise, the Narmada judgement can hardly remain exempt from a process of critical examination. I would go further: it *must* be so examined.

To some, it's likely my preamble will smack of sour grapes. I suppose those of us who object to parts of the judgement will have to live with that.

It was in a spirit of inquiry that I read both the majority and the minority judgements that were handed down on 18 October 2000. Again, I tried to read them not as someone who already believes the dam is a major mistake, but as an ordinary citizen would. I read them with the intent of simply understanding the logic and process of justice as it was handed down in this case.

Here, in no particular order, is some of what I found.

•

The Majority Judgement

Justice B.N. Kirpal wrote the majority judgement, which was co-signed by the then Chief Justice of India, A.S. Anand. It has a number of interesting features. Take first Justice Kirpal's treatment of the Morse Report.

Right from the time work on the SSP began, there were concerns and public protests about R&R, as well as other aspects of the project. For example, as early as August 1983, Arch Vahini, a NGO that has always championed the dam but has steadfastly demanded satisfactory R&R, wrote a strong protest to the World Bank warning of the distress displaced families were facing. In subsequent years, there were more public demonstrations and protests, much of it stemming from the failure to simply inform and consult people who would be affected. By the end of the 1980s, there was widespread opposition to the dam in Madhya Pradesh and Maharashtra; serious concerns about aspects other than R&R were also being aired.

Responding to this situation, the World Bank appointed Bradford Morse to head a commission to review the project in 1991-92. While conducting their review, as I have observed before, Morse and his colleagues had full and generous cooperation from the governments of Gujarat, Madhya Pradesh, Maharashtra and the central government. Again, the Report explicitly acknowledges this cooperation.

Subsequently, the NBA supported its arguments in the Supreme Court in part by referring to the Morse Report.

No doubt the Court—any court—is at liberty to accept or throw out evidence as it sees fit. The Morse Report is no exception. Now I believe the Report's findings are worthy of consideration. Given my feelings about dams, and about the Sardar Sarovar dam in particular, given my opinion of the Morse Report, I would have certainly liked the Court to accept its findings, or at least use them as evidence. Not that I believe everybody must agree with the Morse Report. Certainly, it is possible the Court would have, on reading the Report, found no merit in it.

The vital words in that last sentence, for me, are: *on reading the Report.*

In the majority judgement, Justice B.N. Kirpal observes:

The Government ... did not accept the [Morse] Report and commented adversely on it. In view of the

above, we do not propose, while considering the
petitioners' contentions, to place any reliance on the
report of Morse Committee.

But the Government of India was itself the first respondent
in the case! This positively baffles my non-lawyerly mind.
How can a respondent's view of evidence that is critical of
it be allowed to decide whether a court at all considers
such evidence? If you are a murderer facing trial, and the
prosecution produces the murder weapon as evidence,
would it be acceptable for you to tell the court: 'I don't
like that gun. Please don't consider it as evidence'? More
than that, would it be acceptable for the court to pay heed
to you and throw the gun out of court without examining
it?

I read the majority judgement very carefully, looking
for more tangible reasons the Morse Report was rejected.
I found none. The first respondent's apparent dislike of
the Report was good enough.

And yet, elsewhere in the judgement, Justice Kirpal
quotes approvingly from another report—but an unnamed
one—by the World Bank. He says:

> The cost and benefit of the project were examined by
> the World Bank in 1990 [i.e. before the Morse Report]
> and the following passage speaks for itself:
> [Interjection added]

> 'The argument in favour of the Sardar Sarovar Project
> is that the benefits are so large that they substantially
> outweigh the costs of the immediate human and
> environmental disruption . . .'

In other words, this is a World Bank report that approves
of the dam project.

So from this judgement, we learn that an anonymous
World Bank report that assesses the *potential* of the SSP

and concludes that its benefits will outweigh costs is acceptable. Presumably, also to the Government of India. But another, and more recent, World Bank report of an actual *review* of the project as it stood at the time, a report that severely criticizes the project, is rejected by the Government, and *therefore* ('In view of the above') by the Court.

Remember this revealing sentence (quoted in Chapter 6)—one among many—from the Morse Report:

> The Sardar Sarovar Projects are likely to perpetuate many of the features that the Bank has documented as diminishing the performance of the agricultural sector in India in the past. (Morse: 321)

Is there any reason the Court should treat this excerpt, also from a World Bank report, with any less (or more, for that matter) seriousness than the excerpt from that other one ('The argument in favour of the Sardar Sarovar Project . . .') I quoted some paragraphs earlier?

Second, take the meaning of *pari passu*. This curious little term has been part of the verbiage surrounding the project ever since construction began on the dam. According to the Random House dictionary, it means 'equal pace or progress; side by side.' In giving the project various clearances over the years, authorities have decided that it should meet various requirements *pari passu* with the construction of the dam. Environmental requirements, in particular.

You can argue the correctness of this *pari passu* regime—and I am not the only one who believes it has become an excuse for neglect—but that's a different issue altogether. Still, take an odd reference to it I found in the majority judgement.

Justice Kirpal notes that the project was given

environmental clearance 'subject to certain conditions'. Among those conditions was this one (and here he quotes from a Ministry of Environment and Forests [MoEF] office memorandum of 24 June 1987):

> The catchment area treatment programme and rehabilitation plans be so drawn as to be completed ahead of reservoir filling.

Note: this is *not* a *pari passu* clearance, but a *requirement* that certain tasks must be 'completed ahead' of certain other tasks. The distinction in this case is very clear. Elsewhere in that office memorandum, the Ministry says that 'environmental safeguard measures [must be] planned and implemented *pari passu* with progress of work on [the] project.' But the 'catchment area treatment programme and rehabilitation plans' have to be 'completed *ahead* of reservoir filling.'

Justice Kirpal quotes the entire MoEF office memorandum, and later in his judgement mentions that the NBA's counsel had made reference to it as well. Yet in the paragraph right after this mention, we find him noting:

> The [project authorities] had proceeded on the basis that the requirement [in the Office Memorandum] that catchment area treatment programme and rehabilitation plans be drawn up and completed ahead of reservoir filling would imply that the work was done *pari passu*, as far as catchment area treatment programme is concerned, with the filling of [the] reservoir.

Baffled again, and shades of Orwell as well. How was 'completed ahead' taken to 'imply that the work was done *pari passu*'? The dam-builders are required—not urged or advised or encouraged, but *required*—to complete one thing before doing another, but they simply assume that this *requirement itself* means both things can be done side by side. How did that happen? Does this not violate the clearance—the conditional clearance—given to the project in 1987?

You may think this is a mere quibble. If so, let me ask this question: have the catchment area treatment programme and the rehabilitation plans, in fact, been drawn up and completed? For after all, the reservoir is filling up. Whether 'completed ahead' or done *pari passu*, have those other requirements been met?

Third issue. I've mentioned it in passing here, and no doubt you have heard about it many times, but have you ever wondered what 'environmental clearance' really means? If I were asked, I might say off the top of my head that it means certain mandatory assessments of environmental issues have been done, reports submitted, and all of this been deliberated on. Only at the end of that process would the clearance be either granted or withheld. Either outcome would be equally possible, I imagine, at the start.

Sounds reasonable? Then consider this summary of what happened as the Narmada project authorities sought environmental clearance, as recounted by Justice Kirpal and paraphrased by me.

In January 1980, the GoG submitted a 'detailed project report [on the Sardar Sarovar project] in 14 volumes' to the Central Water Commission (CWC). The CWC sent the report on to the Department of Environment, which wrote back to the Gujarat government asking for information on 12 different 'ecological aspects' of the project. This information took three years to collect, in which time the CWC did a 'techno-economic appraisal of the project . . . and [on 6 January 1983] found it acceptable subject to environmental clearance.'

Over the next few years, the governments of Gujarat, Maharashtra and Madhya Pradesh submitted reports and held several meetings to assess the environmental aspects of the project. All of this was in pursuit of the Department of Environment's requirements. In October 1986, the Ministry of Water Resources (MoWR) sent the MoEF 'a note on environmental aspects . . . and noted the urgency of the decision.' This note listed the tasks that remained to be done and the time each task was expected to take—generally two or three years.

It also considered, wrote Justice Kirpal,

> two options . . . one to postpone the clearance and the other . . . to clear with certain conditions with appropriate monitoring authorities to ensure that the action is taken within the timebound programme. It was concluded that in the light of the position set out, it was necessary that the project should be cleared from the environmental angle, subject to conditions and stipulations outlined.

Whatever that last sentence may mean.

Apparently, as early as 1986, it had become 'necessary' to clear the project. The balance between the two options that I mentioned earlier—granting or withholding clearance—seemed to have tilted already.

But the MoEF still did not produce the clearance. This appears to have weighed on the mind of the MoWR, which eventually saw fit to approach the Prime Minister's Office (PMO). In a note dated 20 November 1986, it again spelled out 'two options with regard to the clearance': one, to wait for two or three years for the completion of all the studies etc; and two, to clear the project right away 'subject to the stipulation with regard to the action to be taken in connection with various environmental aspects and appropriate monitoring arrangements to ensure that the actions were taken in a time bound manner.'

Whatever that last sentence may mean.

In December that year, the MoEF sent a note to the PMO indicating that 'there was *absence and inadequacy* on some important environmental aspects even though the SSP was in a fairly advanced stage of preparedness.' [Emphasis added]. Therefore, the MoEF had some suggestions. Read on to find out what they were.

In January 1987, the additional secretary to the Prime Minister put together a note in turn. Justice Kirpal writes:

The note opened by saying that [the Narmada projects] have been pending approval for a considerable amount of time. The States of MP and Gujarat have been particularly concerned and have been pressing for their clearance. [This] note mentioned that [the MoEF note of December 1986] had recommended conditional approval to the [projects] subject to three conditions:

1. Review of design parameters to examine the feasibility of modifying the height of the dam.

2. Preparation in due time, detailed and satisfactory plans for rehabilitation, catchment area treatment, compensatory afforestation and command area development.

3. Setting up of Narmada Management Authority with adequate power and teeth to ensure environmental management plans are implemented pari passu with engineering and other works.

Fair enough, so far. Despite the appearance here of the *pari passu* principle, the MoEF insists on 'detailed and satisfactory plans' to be produced 'in due time', besides raising the possibility of modifying the height of the dam.

But the additional secretary's note reveals much more. The MoWR and the state governments seem to have 'had no difficulty in accepting conditions 2 and 3.' Fine. But when it came to condition 1, the MoWR had decided that a 'reduction in the dam height did not appear to be feasible' and also 'would not be worthwhile.' (How did the MoWR assume that 'modifying the height of the dam' meant a 'reduction in dam height' and not an increase?) For intangibles such as setting up a body, or producing reports and plans at some unspecified 'due time' in the future, there is 'no difficulty.' But the concrete proposal to modify the height? That is dismissed: it 'would not be worthwhile.'

But exactly why not? What was the problem with reducing the height if the MoEF had chosen to recommend that it be done? Why was it not 'feasible' and not 'worthwhile'? The MoWR was silent.

In any case, writes Justice Kirpal,

it was agreed [by the Secretaries in the PMO, the MoWR and the MoEF] that the recommendation of the [MoEF] of giving clearance on the condition that items 2 and 3 referred to hereinabove be accepted.

(Remember here that the MoEF had recommended clearance to the project subject to *three* conditions being accepted, not just two.)

In fact, the secretary to the Prime Minister wrote a few lines on this note from the additional secretary. She observed:

The [MoEF] have recommended clearance of this project subject to conditions which will take care of PM's apprehensions. I shall request Secretary, Water Resources . . . to see that no violation of any sort takes place. . . . The matter is urgent as last week CM Gujarat had requested for green signal to be given to him before 20th January.

PM may kindly approve.

Stop to give a thought to what was happening in January 1987. The MoEF recommends clearance subject to three conditions. One among those three, the MoWR and the state governments think, 'would not be worthwhile.' So it is simply thrown away. Where you might expect the institution of mechanisms to monitor possible violations, there is a mere expression of intent ('I shall') to 'request' a senior bureaucrat to 'see that no violation of any sort takes place.' To top it all, the CM of Gujarat presses for a decision, a 'green signal . . . before 20th January.'

Why is the matter 'urgent'—only because the CM of Gujarat wants a 'green signal' quickly? The whole spirit of this note is that the project must be cleared because some people—like the CM of Gujarat—want it cleared. Not because it successfully meets certain conditions. There is no sign that any consideration is being given to the possibility of the project *not* being cleared. Or receiving clearance only after requirements are met. The pressure, the *urgency*, of that desired green signal has shot down any possibility of a *red* signal.

Be that as it may, the PM did 'kindly' approve. Or at least, we must assume he approved. As Justice Kirpal notes:

The Prime Minister Shri Rajiv Gandhi, *instead of giving the approval*, made the following note: 'Perhaps this is a good time to try for a River Valley Authority. Discuss.' [Emphasis added]

There, in total, is the PM's 'approval'. There, in total, is the environmental clearance this project was awarded. A clearance that had to come all the way from a prime minister. A clearance that isn't even really one—note Justice Kirpal's use of the word 'instead'. Just so was a major Indian project given its go-ahead, at least on environmental considerations.

But if this strikes you as strange, it should not. There is an explanation that evidently carried some weight when the case was decided. This explanation comes from the Five Member Group (FMG), a panel set up in 1993 to review the project. Here's what Justice Kirpal writes:

> [Gujarat] contended that various alleged dangers relating to environment as shown by the petitioners were mostly based on the recommendations of the Morse Committee Report and Five Member Group. ... [T]he [Morse Report] does not require our attention [because it was rejected by the Government of India]. ... [T]he report of the FMG [reads in part] as follows:

> '... If in spite of all these arrangements [to take care of environmental issues], the environmental point of view fails to be heard adequately, and if project construction tends to take an over-riding precedence, that is a reflection of the relative *political importance* of these two points of view in our system. This can be remedied only in the long term through persuasion and education, and not immediately through institutional arrangements which run counter to the system.' [Emphasis added].

Political importance. Take note.

Blame that amorphous enemy, the 'system', for its inattention to environmental issues. Don't hold officials to their commitments. Above all, allow project construction to go ahead without hitches—not because the merits suggest that that is the thing to be done, but because construction just happens to be *politically* more important than environmental considerations. This is the message in these lines.

Besides, who is really concerned about that 'long term'? Are you? Was the FMG? Is anybody? It is astonishing that while one report—Morse—is again dismissed, the FMG's report is acceptable. This point it makes about 'political importance' and things to be 'remedied only in the long term' is simply repeated, as if it is profundity rather than evasion.

Fourth, take something called the 'precautionary principle'. The petitioners in the Narmada case relied on a 1996 judgement that laid down this principle. It essentially says that in case a proposed project is likely to cause environmental damage, those responsible must prove that such changes will not be harmful.

From that 1996 case (Vellore Citizens' Welfare Forum versus Union of India):

> The precautionary principle suggests that where there is an identifiable risk of serious and irreversible harm [to the environment] it may be appropriate to place the burden of proof on the person or entity proposing the activity that is potentially harmful to the environment.

How does Justice Kirpal react to this principle? As follows:

> It appears to us that the [principle] will ordinarily apply in a case ... where the extent of damage likely to be inflicted is not known. ... The construction of a dam undoubtedly would result in the change of environment but it will not be correct to presume that the construction of a large dam like the Sardar Sarovar will result in ecological disaster. ...

> [T]herefore the [principle] will have no application
> in the present case.

What do we learn here? That the dam will indeed change
the environment. But since we can't presume that it will
be a bad change, the precautionary principle does not
apply. In other words, the precautionary principle applies
except where a judge feels 'it will not be correct to
presume' it will damage the environment. Why then, I
wonder, is it a *precautionary* principle? After all, it protects
against something adverse that might arise out of some
step taken. Is it right to assume *a priori* that there will be
nothing adverse?

Fifth, consider some other miscellaneous points in the
judgement that might suddenly make you sit up a little
straighter.

> [T]he number of PAFs [Project Affected Families] as
> estimated in 1992 by the State Governments were
> 30,144. ... As per the 1990 Master Plan [for
> rehabilitation] the total PAFs have increased to 40,227
> from 30,144 due to addition of 100 more genuine
> PAFs in Maharashtra. ... The reason for increase in
> number of PAFs has been explained in the Master
> Plan and the reasons given, *inter alia*, are:
>
> a. After CWC (Central Water Commission) prepared
> backwater level data, the number of PAFs in MP
> increased by 12,000 PAFs.

Just what does all this mean? In 1992, PAFs were estimated
at 30,144, but in 1990, two years *earlier*, they had 'increased'
from 30,144 to 40,227? And that increase, time-warped
though it appears to be, is because of adding 100 more
Maharashtrian PAFs? But 30,144 + 100 is 30,244, not
40,227. Or is that increase because of adding 12,000
Madhya Pradesh PAFs? But 30,144 + 12,000 is 42,144, not
40,227.

Again, just what does all this arithmetic mean?

If one compares the living conditions of the PAFs in their submerging villages with the rehabilitation packages first provided by the Tribunal's Award and then liberalised by the [three] States, it is obvious that the PAFs had gained substantially after their resettlement.

Which seems to me to be a patently unfair comparison. Let's assume the living conditions of PAFs in their villages were abysmal. To conclude that they have 'gained substantially after their resettlement', surely we should compare those abysmal conditions to their *living conditions* after resettlement? Not to the *packages* spelled out by the Award and the states?

Yes, the packages are filled with good intentions. But let's not confuse their mere existence with reality.

This is not to say that PAFs may not indeed have gained after resettlement, but to make the point that simply producing a glittering package on paper does not amount to such a gain. If that were so, we should, among other things, immediately conclude that all Indian children do have access to free, compulsory primary education. For does our Constitution not contain a Directive Principle of State Policy urging the State to 'endeavour to provide, within a period of ten years from the commencement of the Constitution, for free and compulsory education for children up to fourteen years of age?' Why bother with the reality, which is that over fifty years 'from the commencement of the Constitution', countless millions of our children are not in school and nearly half the country is illiterate? Regardless of that Directive? Why bother with the reality, that a large number of these people displaced by the Sardar Sarovar dam complain about their condition after being displaced? Regardless of the packages?

The majority judgement quotes liberally from a GoG affidavit that lists various measures that government says it is taking with regard to PAFs. Justice Kirpal concludes:

> [T]his Court is satisfied that more than adequate steps are being taken by the State of Gujarat ... and therefore, continued monitoring ... may not be necessary.

So the GoG is simply allowed to state that it is doing a good job—and after all, what else is it likely to tell the Court?—and that is enough to let it off from being monitored.

On the other hand, the NBA, which was the petitioner in this case, is severely criticized. Among other things, Justice Kirpal says the NBA approached the Court far too late and is thus guilty of causing delays and cost overruns in the project. Justice Kirpal is entitled to his view on that score, but in his conclusion, he has this interesting statement:

> In a democratic set up, it is for the elected Government to decide what project should be taken up ... and unless and until it can be proved or shown that there is a blatant illegality in the undertaking of the project or its execution, the Court ought not to interfere with the execution of the project.

Well, just how will a petitioner prove blatant illegality if the Court dismisses his contention because he approached it too late? After all, it was exactly the endeavour of the NBA to show 'blatant illegality in the undertaking of the project', among other things in the whole issue of environmental clearance. The NBA believes the project has never been accorded this approval. That belief is only vindicated by Justice Kirpal's own observation that Prime

Minister Rajiv Gandhi, *'instead of giving the approval*, merely scribbled a note on a memo, and that note was treated as an environmental clearance. The NBA contended that this itself violated guidelines for the project and laws of the land. Whether that constituted 'blatant illegality' was for the Court to decide, which was why the NBA filed the petition in the first place.

Besides, the NBA filed this petition in 1994. Why was it not summarily dismissed at that time if the 'Court ought not to interfere with the execution of the project'? While hearing the petition, this very Court stayed construction on the dam—and therefore did 'interfere with the execution of the project.' So why this concern six years later?

Close to the end of the judgement, Justice Kirpal makes this point:

> The availability of drinking water [from the dam] will benefit about 1.91 lac of people residing in 124 villages in arid and drought prone border areas of Jalore and Barmer districts of Rajasthan who have no other source of water and are suffering grave hardship.

His concern for the sufferers in Rajasthan is appreciated. But in the very next paragraph, we read:

> The *only* benefit from the project which Rajasthan is to get is its share of hydel power. [Emphasis added]
> (Majority judgement: all quotes)

The very next paragraph. At the end of an exhaustive judgement about a dam, there remains ambiguity about what its benefits will be, whom it will benefit.

These questions are precisely what the entire NBA petition was about to begin with.

•

The Minority Judgement

Justice S.P. Bharucha, who succeeded Justice Anand as Chief Justice of India (and was succeeded in turn by Justice Kirpal) wrote the minority judgement. Let me make it clear that his is by no means a judgement that is critical of the project. On the contrary, Justice Bharucha begins by declaring:

> I take the view that the Sardar Sarovar Project *does not require to be re-examined*, having regard to its cost effectiveness or otherwise ... I do not accept the submission on behalf of the petitioner that those ousted by reason of the canals emanating from the reservoir in the Project must have the same relief and rehabilitation benefits as those ousted on account of the reservoir itself; this is for the reason that the two fall in different classes. [Emphasis added]

Elsewhere in his judgement, he is even more categorical:

> Despite the submergence of land and displacement of people and livestock, there was *no case for the abandonment* of the projects. [Emphasis added]

Yet in a verdict that is several times shorter than Justice Kirpal's, Justice Bharucha examines very closely various aspects of the project's environmental clearance and rehabilitation. His examination leads him to rule, among other things, that:

> Until environmental clearance to the Project is accorded by the Committee of Experts [that he says must be set up], further construction work on the dam shall cease.

On the same evidence presented to the other Justices on the bench, Justice Bharucha has reached diametrically opposite conclusions, among them that this project *never did get* its environmental clearance.

But let's look a little more closely at this minority verdict. Consider first the very same example of the use of *pari passu* that I excerpted from the majority judgement above. Here is Justice Bharucha's paragraph on this point:

> While environmental safeguard measures were to be planned and implemented *pari passu* with the progress of the work on the project, the catchment area treatment programme and the rehabilitation plans were required to be 'so drawn as to be completed ahead of reservoir filling.' This condition *clearly required* that before any water was impounded in the reservoir the catchment area treatment programme was not only to be drawn up but also *to be completed*; so also the rehabilitation plans. If as the Project authorities interpreted this clause, only the drawing of the catchment area treatment programme and the rehabilitation plans were to be completed ahead of reservoir filling, the clause would have read: 'The catchment area treatment programme and rehabilitation plans shall be drawn ahead of reservoir filling.' What the clause as drawn required was that the catchment area treatment and the rehabilitation works would be completed ahead of impoundment in the reservoir. This, plainly, was intended to offset, so far as was possible in the circumstances, the adverse effect of the impoundment of water in the reservoir upon the catchment and those who were required to be settled elsewhere. In fact, the impoundment began much before. [Emphasis added]

With no room for uncertainty, Justice Bharucha criticizes the project authorities' assumption that the requirement 'implied' that they could do certain things *pari passu*. And what's worse, his judgement also points out that even the assumed *pari passu* scheme was ignored: for 'in fact, the impoundment began much before.'

In other words, the only thing the dam-builders were enthusiastic about was to proceed with the construction of the dam and filling its reservoir. Everything else could be safely ignored. But while Justice Bharucha rules that the Project authorities must be held to commitments they make, his colleagues on the bench took no note of aspects the authorities ignored.

There was also Justice Bharucha's own examination of the environmental clearance given to the project. In one sentence, he picked up the fundamental flaw in this clearance:

> It appears that, though it ought rightly to have been taken by the Ministry of Environment and Forests, the decision whether or not to accord environmental clearance to the Project was left to the Prime Minister.

Emphasizing a few points the October 1986 MoWR note makes, Justice Bharucha writes:

> Under the sub-heading, 'What still remains to be done', the [MoWR] note stated, 'While some plans have been made, studies undertaken and action initiated, it will be clear . . . that much still remains to be done. Indeed, it is the view of the [MoEF] that what has been done so far whether by way of action or by way of studies *does not amount to much*, and that many matters are as yet in the early and preliminary stages.' [Emphasis added]

'Does not amount to much.' Could there be a more explicit declaration of the apathy of project authorities towards environmental issues?

That MoWR note said that one possibility for the project was to postpone the environmental clearance by three years, giving time for completing the various surveys and studies that had to be done. But here's how the MoWR viewed such a postponement:

> With the [clearance] postponed for three years, and with no assurance that at the end of that period, the decision will be positive, it is difficult to believe that all these studies, surveys and plans relating to the environmental aspects will be pursued with energy and enthusiasm, and the necessary resources devoted to them.

Now there's finely-crafted logic for you. Because studies haven't been done, clearance for the project should be postponed to allow time for them to be carried out. But since it is by no means sure that the clearance will, in fact, be accorded after those three years, people will be half-hearted about these studies. Therefore, it's better to give the clearance right now. In other words, the clearance is going to be awarded in any case. Therefore why wait three years? Why not award it now? Why fool around with trivialities like studies?

It was Ram Manohar Lohia, a man who helped frame our Constitution, who once made a suggestion that falls right in line with this MoWR logic. He proposed that at birth, every Indian should be awarded a university degree. An elegant proposal indeed. After all, at birth and for a good twenty years thereafter, every Indian has 'no assurance' that he will acquire a university degree.

Therefore, we can—must—assume he is unlikely to pursue any of his studies through school and university 'with energy and enthusiasm.' So why not give him the degree at birth?

Or take another analogy. You enter an essay contest. Is it reasonable to tell the organizers at the start: 'If you do not assure me of the prize, I will not be able to write my essay with energy and enthusiasm. That is, if you do not assure me of the prize, I cannot produce an essay worthy of winning the prize. So please give me the prize right now.'

And anyway, do I really need to point out that those studies and surveys the MoWR refers to have hardly been 'pursued with energy and enthusiasm' even after the clearance was awarded?

About the MoWR's note to the PMO of November 1986, Justice Bharucha comments:

> Considering the magnitude of rehabilitation, involving a large percentage of tribals, loss of extensive forest area rich in biological diversity, enormous environmental cost of the project and considering the fact that the basic data on vital aspects was still not available, 'there could be but one conclusion, that the project(s) are not ready for approval.'

An empty conclusion by the MoWR, really: as Justice Bharucha himself points out, this very note recommends that the project be *given* the approval (also see p. 170 of this book).

A December 1986 MoEF note to the PMO ran through laments that were much the same. It said:

> [T]he absence and inadequacy of data on important environmental aspects still persists.

> Even though SSP [Sardar Sarovar Project] is in a
> fairly advanced stage of preparedness, it is neither
> desirable nor recommended that the SSP should be
> given approval in isolation on technical and other
> grounds.

Do you wonder as I do: despite all these words written
about *not* giving the project approval, how was it passed
anyway? But, of course, we know what kind of approval
that was (see Rajiv Gandhi's note, p. 173) and Justice
Bharucha refers to it as well. Further, he comments:

> Clearly . . . the necessary particulars in regard to the
> environmental impact of the Project, as required by
> [MoEF] Guidelines, were not available when the
> environmental clearance was given, and it, therefore,
> *could not have been given.* . . .

> Under its own policy, as indicated by the Guidelines,
> the Union of India was bound to give environmental
> clearance *only after* a) all the necessary data in respect
> of the environmental impact of the Project has been
> collected and assessed; b) the assessment showed that
> the Project could proceed; and c) the environmental
> safeguard measures, and their costs, had been worked
> out. [Emphasis added]

Aside: since the project went ahead without getting its
environmental clearance, may we conclude that there is
that 'illegality in the undertaking of the project' that the
majority judgement spoke of (see p. 178)? Or at least that
there was reason enough to ask a Court to examine this
issue to determine if there had been any illegality?

Justice Bharucha concludes:

> This Court cannot place its seal of approval on so vast
> an undertaking as the Project without first ensuring

that those best fitted to do so have had the opportunity of gathering all necessary data on the environmental impact of the project and assessing it. ... Until environmental clearance to the Project is accorded by them, further construction work on the dam shall cease. (Minority judgement: all quotes)

'Cease' is not an accurate description of what has happened to construction work on the Sardar Sarovar dam.

10

The Road Less-travelled

I have been walking a sort of tightrope while writing this book. On the one hand, it must be clear that I think dams lead to many problems; I also think the dams on the Narmada, the Sardar Sarovar in particular, are mistakes. On the other hand, I cannot deny that any number of human enterprises are wrongly undertaken and bring problems in their wake. We do persist with them anyway.

Besides, with dams there's a point at which the arguments for and against them end in stalemate. It is impossible for an otherwise disinterested person to decide which side's case has greater merit. Therefore, I don't think that the problems with dams, by themselves—siltation, salination and so forth—are the reasons to cite to stop building Sardar Sarovar. Or at least, they are not arguments that will convince enough people that the dam is a mistake.

In this book, I've attempted to persuade you that the way this dam is pursued—the way projects like these in general are pursued—is itself the case against the dam. Among other things, the debate over large versus small dams is essentially irrelevant to this case. For I am convinced even small dams would have been built in the same slipshod way. It is the process of building dams, of development itself, that is flawed. Not so much the *fruits* of development—dams, power plants or others—but the *process* itself. For it has left so many people so destitute and resentful that they have no option but to complain.

Yet the same process of development also decides that these complaints must not be heard. Or if they are heard, they must be ignored as the rantings of recalcitrant Luddites trying to hold up some near-mythical, even mystical, 'progress' of the country.

That is, if they are not simply branded anti-nationals.

So given all that, where do we go from here? How can we ensure that future 'development' happens fairly, not shoddily? How can we ensure that every national enterprise does not wind up in an acrimonious dispute that the Supreme Court needs to adjudicate; that every such resolution does not leave one or both sides feeling dissatisfied and angry?

•

Regardless of the Supreme Court decision, some compromise in this project is an imperative. Its opponents are determined to keep up their opposition because they believe they have excellent reasons to do so; the builders are as determined to see the project through because they believe the dam must be built. These two positions are so utterly disparate as to be irreconcilable. Without any compromise, the disputes will continue and there will only be further recourse to the courts, or to some extra-judicial authority.

To me, if we are looking for a compromise, the Paranjape-Joy proposal that I explained briefly in Chapter

4 is as good as any, better than most. Or at least, it spells out the basic form that such a compromise must take: stop building the dam at a point where it will deliver most of its claimed benefits. What's more, give two aspects of the project top priority: R&R for the people displaced, and delivery of those claimed benefits to the beneficiaries who most need it.

Like all good compromises, this one will leave both sides contented as well as dissatisfied. The project will not be totally abandoned, which will disappoint its opponents, but gladden its proponents. Yet it will proceed till the dam has reached a height that's a good thirty metres lower than originally planned. This will annoy proponents, but please opponents. And with that height, as Paranjape and Joy spell out, and with a change in the way water is seen and used, Gujarat, in particular, will face no reduction in the benefits it expects from this dam. What's more, Kutch and Saurashtra will get water far sooner than currently planned, and substantially fewer people will face displacement.

All of this seems so profoundly reasonable that you wonder why the Paranjape-Joy proposal has been, and remains, so ignored. Even if this is not the ideal compromise, another can be worked out. The important thing is to recognize that there is a dispute, and so be willing to consider and fully implement a solution, a compromise, that is fair to all.

Today, the continuing construction of the dam is manifestly causing a great deal of resentment and dissatisfaction, some of which you have read about in this book. Yet abandoning it now will certainly also cause a great deal of unhappiness to many people. Like Lirabhai, in his village in Kutch. But these are the only two options—continue or abandon?—that an appeal for adjudication by the Supreme Court make available. On the other hand, meetings and negotiations between people on either side of the dam might produce the compromise this project has never seen.

Maybe Lirabhai and the farmers from Nimad must meet and exchange thoughts more often. Maybe many more such meetings must happen.

•

But what about the larger issue that I wanted this book to address: the half-hearted way in which this dam has been built? What's the guarantee that any compromise will proceed any more efficiently?

Answer: none. This is what I meant when I alluded elsewhere to the state of our roads, among other things. Really, why should anyone have any confidence in the ability of authorities to build anything: dams, road or other projects?

And yet, it is hardly my case that all so-called 'development' must grind to a halt merely because it has been carried out so shoddily in the past. The state must build roads, bridges, buildings, even dams; it has to generate electricity and take water for drinking, irrigation and recharge of groundwater reserves to areas where it is needed. None of this can stop.

But the lesson from the Sardar Sarovar experience must be absorbed and learned. Project authorities must be willing to be transparent in their operations, make every aspect of their work open to public review and criticism. They must not get away, any longer, with wrapping projects in layers of jargon and numbers that conceal far more than they reveal. This public scrutiny will itself promote the attention to detail, the excellence, the fairness, that are lacking in so many aspects of the Narmada dam projects.

More than that, these projects are meant to benefit thousands, even millions, of ordinary citizens. Why should those citizens not know how they are proceeding? For that reason alone, they must be carried out in full public view.

That's a long list of expectations, I know. How will these things ever happen? To begin, at least, I would like to see the press pore through glossy publications put out by agencies like the SSNNL, and critically examine the various claims and figures in them. A couple of discrepancies can be overlooked. But certainly not substantial differences on the most basic aspects of the project. They speak of a broader shabbiness that can only call the entire project into question.

●

What about the arguments of those who oppose the dam, and, of course, I particularly mean the NBA here? I have deliberately kept them largely out of this book, because I want to show that the case against the Narmada dams stands irrespective of the NBA's existence.

Yet, of course, any discussion of the Narmada dams cannot—must not—leave out the NBA. When I consider the NBA's opposition to Sardar Sarovar, which I believe is founded on clear-headed logic, I wonder: why have they not succeeded in convincing more people than they have of their case?

Somewhere along the line, I believe the NBA made a tactical mistake that has left them painted into a corner. They have allowed the debate to become an argument over dams themselves, and specifically the large versus small issue.

To outsiders, this is how their argument must sound: given all the known problems with large dams—siltation, waterlogging, the number of oustees and the vast problem of rehabilitating them, and more—this dam, and all large dams in the country, must be stopped. We must find other ways to deliver water and generate power.

Yes, we must find other ways to deliver them. But I still think the stand against large dams—even though it makes sense to me—was a mistake, and on two counts.

First, the authorities responded to cited problems with pronouncements that they would 'take care' of waterlogging and salinity, or that far more people will benefit than are being displaced, or that R&R is proceeding satisfactorily. Recall Justice Kirpal quoting a Gujarat government affidavit on R&R and then noting that he 'is satisfied that more than adequate steps are being taken by the State of Gujarat.'

Whom do you then believe? In fact, there's no debate any more: the NBA points out some problem, the dam-builders say they are, or will be, addressing it, and that's it. Outside observers nod and say, 'Fine, the problems are being taken care of, so what's the difficulty?' The NBA loses the public opinion battle—which is the one that counts, far more than Supreme Court decisions do—right here. Thus it is that the NBA is easily painted as blindly opposed to the progress of the country, famously obstructionist. This is easier to do than to pay serious attention to the compelling issues the NBA raises.

Second, and this is the real problem, the focus on the evil of large dams—a focus that will never persuade enough people anyway—deflects attention from the way officialdom builds dams. That, and not the wrangling over what dams do, is what truly inspires profound doubt in the ability of Sardar Sarovar to deliver on its promise. For consider this: if the project—and indeed all such projects—had been conscientiously pursued right from the start; if R&R had been addressed with sympathy and dispatch; if required studies had been done when they should have been, and so on; if those things had happened, there would have been far less scepticism about the building of the dam in the first place.

It is because these issues have been addressed so shabbily that there is controversy at all. This is itself the loudest argument against Sardar Sarovar. Yet my feeling is that it has never been adequately examined.

As things stand today, the NBA finds itself shouting about the issues it cares for, but so few listen. I think this deafness has come about because the NBA has become associated, fairly or not, with a blanket opposition to dams, and that opposition is not convincing enough to many people. They believe that dams are necessary; of the others, some believe that while some or most dams are unnecessary, this one is needed; still others believe that even recognizing the problems with dams, dams like Sardar Sarovar are the *only way* to take water to parched areas.

It's hard to persuade people who speak of the *only way*.

Therefore, to win over public opinion, the dam must be fought from within these lines of arguments, within the case its proponents make for it. Plans incomplete, clearances given hastily, if at all, figures fudged, progress wildly overstated, on and on—these are the hallmarks of this massive project. The seeds of its failure are sown here.

If the NBA had relentlessly brought these issues to public notice, I believe it would have found a much greater level of support—and crucially, in Gujarat as well. With that, it would have been much harder for the builders of the dam to go ahead with it. Instead, the NBA is caught between the rock of implacable official and Gujarati hostility, and the hard place of widespread public indifference to its moves, suspicion of its motives.

It need not have been this way.

•

One of the triggers for this book was the Supreme Court judgement. When I sat down to read it, I felt some of the same desolation that struck me when I first heard about it, that October day. But it was hardly a desolation that came from disagreement: after all, I read it knowing that I would disagree. What troubled me were the twists and turns it went through to justify this project; the leaps of illogic that littered the route to the decision; the willingness of the Court to accept pretty much every explanation or excuse from various authorities. The reluctance to hold those authorities to their commitments.

All that troubled me because it spoke of the same approach to development that I have alluded to throughout this book. A sort of—to borrow a phrase—cavalier attitude to complex problems. An indifference to humanity.

I find it frightening that I can say all that about our judiciary.

I find it frightening that this is where development has brought us.

Appendix A

More From the Horse's Mouth

In the agenda for the fiftieth meeting of the Narmada Control Authority's Sub-Group on R&R, held on 29 August 2001 in Bhopal, tucked away in Annexure L-7, there's this revealing paragraph:

> Based on physical verification, the land area to be affected [by the dam's reservoir] now works out to 23,425 ha ... The earlier area of submergence, 20,722 ha, was based on simply drawing of contour lines on the village maps. The difference between earlier appraisal and the estimate made now is 2,703 ha, i.e. an increase by 12.75 per cent.

Just like that, buried deep in a mass of typewritten pages, we have an admission that 12.75 per cent more land is going to vanish than earlier estimated, with all the human costs that is likely to entail. Not that a 12.75 per cent discrepancy is a vast one; but if 'simply drawing of contour lines on the village maps' was an unreliable method, why was it used in the appraisal? Are we really being told that when the plans for this project were being drawn up, the authorities did not even have accurate maps, or know how to use them to correctly estimate the area of submergence?

When R&R is already such a problem, what are the prospects for the additional people to be displaced because of this 'physical verification'?

•

In his essay about the dam in a SSNNL booklet, the ex-chief minister of Gujarat, Babubhai Jashbhai Patel, makes this claim:

> [T]he flow of 28 MAF [in the river] is estimated on the observation of 79 years between 1891-92 and 1969-70 ... It is not a mere guesswork [sic], but an assumption based on observed data of 79 years. (SSNNL: 20)

A firm pronouncement indeed? But as you have already seen (Chapter 1), it is quite wrong. First, 28 MAF was the amount the four states concerned *agreed among themselves* should be *assumed*, irrespective of observations. Second, the Tribunal's own calculations using the data for those years was that the flow was 22.6 MAF; it then ignored this figure and went along with the states' desire to assume it as 28 MAF. Third, there were no actual observations of the flow for the years between 1891 and 1948: data for these years was *hindcasted* based on the relationship between rainfall and flow that available data for 1948-70 suggested. Not 'observed data' of water flowing in the river, but an extrapolation from existing *rainfall* data.

The truth is, it's not far wrong to say that the 28 MAF figure was indeed 'a mere guesswork'.

•

It also turns out that the figures in the *FACTS* booklets provide ample reason to doubt the extent to which project authorities give careful attention to the benefits from the dam. Take a subtle example. On page 26 of the 1998 *FACTS* is this sentence: 'About 75 per cent of the command area in Gujarat is drought prone.' Thirty pages later, on page 57, this claim now reads: '[T]he project is designed to benefit about 75 per cent of the drought prone area . . . of the State.'

These sentences appear to be saying the same thing, but they are not. They only underline the carelessness with figures that permeates these publications.

•

Too often in the SSNNL booklets, figures are simply tossed about, statements simply made, as if their presence on the page alone vouches for the excellence of the project. In his booklet for the SSNNL, B.K. Jhala devotes a couple of pages to addressing the environmental benefits of the 'proposed large scale afforestation plan' that's part of the project. We learn from these paragraphs that the authorities say they will plant a total of *over 12 million* trees in and around the dam. This is a number too big to really mean much, but it looks impressive.

Still, let's assume that project authorities will indeed plant that many trees. There are two interesting points that follow this claim in Jhala's booklet:

> For protecting the fauna, the proposed area of the sanctuary has been increased from 150 sq kms to 445 sq kms. (Jhala: 6)

Good. But which sanctuary is this? This is the first mention of any sanctuary in the entire publication, and nowhere else does he specify which sanctuary he means.

> Although no effective solution for the prevention of air pollution has yet been found out, but according to an experiment conducted in Germany, it has been established that a tree of the age range 40-50 years has a capacity to suck 2,300 Kgs of Carbon Dioxide and emit 1,700 Kgs of Oxygen. Thus, if trees are

planted in 50 metre broad strip around a chemical factory, they would reduce 90 per cent of the Sulphur Dioxide and 67 per cent of the Nitric Acid . . . [and also] reduce 20 to 30 per cent of noise pollution. (Jhala: 6-7)

How many questions does this paragraph raise in your mind? I wonder about at least these: Which experiment is this and when was it 'conducted'? In fact, just where was it conducted, I mean more precisely than a mere 'Germany'? In what stretch of time does the tree 'suck 2,300 Kgs of Carbon Dioxide and emit 1,700 Kgs of Oxygen'? An hour? A week? A year? Its (remaining) lifetime? How does sucking carbon dioxide and emitting oxygen 'reduce 90 per cent of the Sulphur Dioxide and 67 per cent of the Nitric Acid', let alone '20 to 30 per cent of noise pollution'?

How are we to know the meaning, the truth, of all this, apart from just being told it blandly in this booklet? Besides, these goodies come from trees that reach the age of forty or fifty years. Yet again, the SSNNL tells us about 'benefits' of its project at an effectively meaningless date in the future. How do we even know these trees will, if they are planted at all, last forty or fifty years?

Want a scientific, closely argued reason to build this dam? I found one in the 1998 *FACTS*, at the end of the discussion about droughts in Gujarat in the mid-1980s. 'The drought is such a menace,' it says, that

Even the World Bank has reported in one of its findings that if water is not taken of [sic] these areas, then by the year 2021 people of these areas would come on the streets. These are the major factors in considerations [sic] which have lead [sic] to the decision of implementing SSP. (*FACTS* 1998: 15)

Let me ask again: How many questions does this paragraph raise in your mind? I wonder about at least these: What 'finding' of the World Bank is this? May we have its full bibliographic reference? What is the meaning of saying people 'would come on the streets', and what 'finding', World Bank or otherwise, would make such a bland, vague pronouncement? In any case, why would the 'people of these areas' wait till 2021 to 'come on the streets'? Why not now, when water is so scarce already? But mere mention of 'World Bank', like an earlier mention of 'Germany', is assumed to be enough to persuade critics of the boons of this project.

Nor is this an isolated instance. Later in the same booklet there's this sentence: 'The World Bank it self [sic] has acknowledged the world class technology in this project. It is pardoxial [sic] to consider the assumptions as suspect.' (*FACTS* 1998: 56)

Yet again: when and where did the World Bank issue this acknowledgement? Why is it that the very way the Bank is mentioned here, the tone that says 'Don't argue, after all the Bank has approved of our work', makes me 'consider the assumptions as suspect'?

What's more, a few lines further down the page is this nearly identical sentence: 'The morse [sic] commission [sic] itself has acknowledged the world class technology in this project.' But I also recall that the Government of India explicitly rejected the Morse Commission Report, which rejection, you will remember, the Supreme Court used to justify its refusal to consider the Report as evidence. In that case, why should the Morse Commission's acknowledgement, if there really was one such, be trumpeted here? In fact, since the government rejected the report, why was this acknowledgement not treated as a black mark against the technology in this project?

Appendix B

More Government Responses to Criticism

1: *The Government of Gujarat's response (GoGR) to the Morse Report (IR)*

Throughout, the authors of the GoGR are intent on giving their readers the impression that the IR is wrong and incomplete and so forth. So on page 18, they tell us: '[I]t is important to repeat that there are a very large number of factual errors in the Independent Review's chapter on R&R.' Well, fine. But what are these errors? 'We discuss such instances—each one briefly—in the next seven pages,' they write. But the next seven pages hardly point out a slew of *errors*; they discuss various points of disagreement with the IR.

A subtle distinction, you think? But think about what it means to say that a Report has 'a very large number of factual errors.' Think about the impression of that Report such a statement puts in your mind.

Anyway, here are just a few points from those seven pages.

On page 15, the GoGR draws a distinction between Section 144 of the Criminal Procedure Code and the Official Secrets Act. 'The two are not related with each other,' it says. This point, because the IR 'refers to the area [the dam site] being placed under Section 144 of the Official Secrets Act, whatever that means.'

Undoubtedly, the IR has blundered here (on page 91, to be precise). But what is the point it is making? That 'no more than five people could gather together for political meetings'. That Section 144 was used against villagers who 'took part in demonstrations as part of the campaign that began in 1988 to secure oustee status', an estimated 200 of whom had been charged under that Section by the time the Review was done.

Why this treatment of these demonstrators? Why was this curbing of protest demonstrations necessary? Are these questions answered by gloating over the IR's unfamiliarity with the precise titles and Sections of our laws? Nowhere does the GoGR address the questions; gloating is apparently enough.

On the same page: 'It is incorrect to imply that the affected persons in the six villages have not met the CM . . .' (Independent Review, p. 93)

But on page 93, the IR quotes a woman from the village of Kothie saying: 'We tried to talk to the CM, but we did not get to see him.' That is, this isn't the IR *implying* that some people could not meet the chief minister, it is one affected woman *saying* she could not do so. Should the IR not listen to people like her? Not quote what they say? If she says she could not meet him, is the IR 'implying', and 'incorrectly' at that, that they have not met him?

'Declaring persons "oustees" after 20 or more years,' says the GoGR (still on page 15), 'can create obvious legal and practical problems. The way out is to implement quickly a reasonable compensation package to solve this problem.' This is about the villagers of Kevadia.

The IR says (still on page 93) that the Kevadia villagers 'fall within the unambiguous language of the . . . agreement's definition of oustees displaced from their "usual habitat due to the carrying out of the Project." '

That is, they are oustees by the *Government of Gujarat's own definitions*. If there are 'obvious legal and practical problems' with applying those definitions, they must be dealt with. The mere possibility that these problems will erupt is no reason *not* to treat these villagers as oustees. Which is just the argument the IR makes.

And in any case, the IR ends this section by calling 'for a quick and generous solution to the problem.' Not very different language from the GoGR's stated desire to 'implement quickly a reasonable compensation package to solve this problem.'

Page 20 begins with these sentences: 'The NWDT Award provisions quoted on page 132 [of the IR] is not correct. Nowhere in the Tribunal Award is there a mention of Gujarat villages whose rehabilitation is to be completed by 1983.'

The trouble is that nowhere on page 132 of the IR is there a mention of any such NWDT provisions, or a 'mention of Gujarat villages whose rehabilitation is to be completed by 1983.' (The figure '1983' does not appear on the page). The sole reference to the Tribunal reads like this: '[T]he Tribunal's 1979 award established that resettlement of a village had to be completed a minimum of six months before its first inundation.'

Wondering if the GoGR had inadvertently got the page number wrong, I searched on several nearby pages for such a reference, and also on such pages as 122, 142, 13, 32 and so on. Nothing. What then is the meaning of this claim in the GoGR? Was it inserted solely to vilify the IR?

On page 21, the GoGR tells us: 'The Review has commented on the lack of irrigation facilities at this site [Gadkoi]. The entire piece of land purchased is covered under irrigation.'

Maybe it is, but what is the relevant portion of the IR? Page 98, on which you read that '[The Gadkoi villagers] also told us that out of their 35 acres between 5 and 10 are irrigated.' Again, should the IR not report what villagers say? Does such reporting amount to 'commentary'?

Finally, what is the 'core problem' with R&R, as the IR identifies it? That R&R measures 'have been effected *pari passu*' with the construction of the dam (page 134, in the conclusion of its examination of R&R in Gujarat). The GoGR is silent about this core problem.

'No one can deny that Gujarat has achieved successes in implementation of its policies,' says the IR. But families who have got irrigated land, it also says, '[seem] to constitute a minority of the total oustee population in Gujarat. When it comes to the prospects of others affected by the Projects . . . serious doubts must arise.'

Nothing in the GoGR is able to explain away those doubts.

Table 6.3 Number of agricultural plots allotted at different dates (IR: 101)

Village	Agricultural Plots Allotted Up To			Recognized Oustees
	March 1989	*Feb 1990*	*Feb 1991*	*March 1991*
Panchmuli	68	183	219	366
Khalvani	24	37	42	110
Navagam	20	24	37	145
Limdi	55	89	144	272
Zer	10	26	24	36
Total	**177**	**359**	**466**	**929**

(Sources: Centre for Social Studies, Surat; Monitoring and Evaluation Report Nos 10 & 12; Nigam monthly resettlement reports.)

2. The Daud Committee Report: Bureaucrats' Dissenting Notes

Reacting to the Daud Committee's observation that the resettlement sites they visited lacked irrigation, the Irrigation department's chief engineer and joint secretary, D.M. More, says: '[T]he process of providing irrigation devices . . . is in progress. The completion of this will take some time.'

Fine, but the process of building the dam and displacing people is also in progress. Why are those tasks not approached as lackadaisically?

More also wrote:

> One has to appreciate that the irrigation facility is to
> be provided within the limits of economic constraints
> and physical availability of water resource. The
> Government is indeed committed to explore all the
> possibilities in this regard to make this activity more
> advantageous with a view to improve the economic
> conditions of the affected tribals as envisaged in the
> NWDT Award. (Daud: 62)

What does this mean? Strictly nothing. If the Government is 'indeed committed', why the complaints that the Daud Committee recorded? And is it reasonable to respond to those complaints by merely reiterating that the Government is 'indeed committed'?

But More also admits that 'during the course of visits I came across a large number of grievances for non-payment of house compensation.' To address this, he proposes to institute a 'small group' that will conduct a 'door-to-door survey of all the affected villages and arrive at the factual details in respect of both the collective and individual entitlements.'

Good point. I'm all for it, as were More's colleagues on the committee. (Except, as you will see, for the R&FD

officials). Only, it strikes me that a small group like this will be just one more in the long list of committees and groups that have looked at different aspects of the dam. Where are all their surveys and reports and 'factual details', and what effect have they had on the dam and its builders?

Besides the recourse to 'should' that I discussed in Chapter 7 (p. 137), the three R&FD men on Daud's panel (Principal Secretary Nand Lal, Section Officer K.S. Parab and Joint Secretary D.R. Mali) also fill their report with a detailed explanation of Maharashtra's 'more liberal package [of R&R] to the oustees of the SSPAFs in Maharashtra.'

This is that old tactic at work all over again. Faced with criticism of policy that is poorly implemented, if at all, they simply spell out the policy at great length, to show how good it really is. Of course it's a good package, nobody denies that. The problem is that it is not being implemented! (Perhaps it *should* be implemented).

Bibliography

Amin, Nanubhai and Naik, M.N., (no date), *Sardar Sarovar Project: A Realistic Perspective*, Gujarat State Committee of the World Wide Fund for Nature—India, Baroda.

Bharucha, Justice S.P., (2000), Minority Judgement in W.P. No. 319 of 1994. Supreme Court of India, New Delhi, October.

Bhatt, Sheela, (2000), 'The Dam . . . and its damned politics (Rediff Diary),' *rediff.com*, Bombay, 2 November.

Bissell, Richard E.; Singh, Shekhar; and Warth, Hermann, (2000), Maheshwar Hydroelectric Project: Resettlement and Rehabilitation, Report of an Independent Review conducted for the Ministry of Economic Cooperation and Development, Government of Germany, reprinted by Narmada Bachao Andolan, Mandleshwar, June.

Black, Maggie, (2001), 'Narmada Dams (cover story),' *New Internationalist*, July.

Comment on the Report of the Independent Review Mission on Sardar Sarovar Project, (1992), Government of Gujarat, Gandhinagar, August.

Dalal, Lalit, (1989), *Namami Devi Narmade*, Sardar Sarovar Narmada Nigam Limited, Gandhinagar.

Datye, K.R., (1997), *Banking on Biomass: A New Strategy for Sustainable Prosperity Based on Renewable Energy and Dispersed Industrialism*, Centre for Environment Education, Ahmedabad.

Daud, Justice S.M., (2001), The Report of the Chairman and other Non-official Members/Invitee, Committee to Assist the Resettlement and Rehabilitation of the Sardar Sarovar Project-Affected Persons (Government of Maharashtra), Mumbai, June.

D'Souza, Dilip, (1996), 'You Need A Thorn To Remove a Thorn,' *The International Indian Woman*, Dubai, March.

D'Souza, Rohan, (2001), 'The Price of Development,' the *Telegraph*, Calcutta, 25 July.

Dubey, K.C., (1987), *A Plan for Roof: Bargi Dam Project*, Office of the Commissioner, Jabalpur, Madhya Pradesh, February.

Dworkin, Ronald, (2001), 'A Badly Flawed Election,' *The New York Review of Books*, New York, 11 January, pp 53-55.

'Gujarat's Narmada Minister resigns,' (2000), *rediff.com*, Bombay, 20 December.

Hari, Johann, (2001), Interview with Gore Vidal, *The New Statesman*, 15 October.

Human Development Report 1996, (1996), United Nations Development Programme, Oxford University Press, New York.

Shariff, Abusaleh, India: Human Development Report, (1999), National Council of Applied Economic Research, Oxford University Press, New Delhi.

Jain, L.C., (2001), *Dam vs Drinking Water: Exploring the Narmada Judgement*, Parisar, Pune.

Jhala, B.K., (1988), Sardar Sarovar Project & Environment, Sardar Sarovar Narmada Nigam Limited, Gandhinagar.

Kirpal, Justice B.N., (2000), Majority Judgement in W.P. No. 319 of 1994. Supreme Court of India, New Delhi, October.

Kothari, Ashish, (2001), 'Against a People's Movement,' *Frontline*, Chennai, 21 July-3 August.

Kumar, Krishna, (1995), 'Learning about Narmada: More Dreams for Politicians' Grandsons,' the *Times of India* (edit page), 25 May.

McCully, Patrick, (1998), *Silenced Rivers: The Ecology and Politics of Large Dams*, Orient Longman, New Delhi.

Mehta, Sanat, (1988), Sardar Sarovar Project: A planned ecological harmony amongst Men, Water, Land & Vegitation [sic], Sardar Sarovar Narmada Nigam Limited, Gandhinagar, October.

Mehta, Sanat, (no date), Interview by Janmabhumi, in Sardar Sarovar Project: The Lifeline of Gujarat, Sardar Sarovar Narmada Nigam Limited, Gandhinagar.

'Minister calls CM liar, gets the boot,' (2000), *Hindustan Times*, New Delhi, 21 December.

Morse, Bradford, and Berger, Thomas, (1992), Sardar Sarovar: Report of the Independent Review [The Morse Report], Resource Futures International, Ottawa.

Paranjape, Suhas and Joy, K.J., (1995), *Sustainable Technology: Making the Sardar Sarovar Project Viable*, Centre for Environment Education, Ahmedabad.

Paranjpye, Vijay, (1989), *The Narmada Valley Projects: A Holistic Evaluation of the Sardar Sarovar and Indira (Narmada) Sagar Dams*, INTACH, New Delhi, September.

Patel, Babubhai Jashbhai, (1988), 'Urgency for Harnessing Narmada,' in Sardar Sarovar Project: The Lifeline of Gujarat, Sardar Sarovar Narmada Nigam Limited, Gandhinagar.

Patel, Jashbhai, (1995), *The Myths Exploded: The Unscientific Ways of Big Dams and Narmada Project*, EDSA, Bombay, August.

Patel, V.B., (1988), Sardar Sarovar Project: A Ray of Hope. Sardar Sarovar Narmada Nigam Limited, Gandhinagar, October.

Pillsbury, Arthur F., (1981), 'The Salinity of Rivers,' *Scientific American*, July.

Raj, P.A., (1998, 1989), *FACTS*: Sardar Sarovar Project. Sardar Sarovar Narmada Nigam Limited, Gandhinagar.

Rehabilitation & Resettlement: Sardar Sarovar Project, (no author, no date). Sardar Sarovar Narmada Nigam Limited, Gandhinagar.

Reisner, Marc, (1986), *Cadillac Desert: The American Desert and its Disappearing Water*, Penguin Books, New York.

Roy, Arundhati, (2000), 'The People vs The God of Big Dams,' the *Times of India*, 25 October.

Sachs, Wolfgang, ed., (1992), *The Development Dictionary: A Guide to Knowledge as Power*, Orient Longman, New Delhi.

Sardar Sarovar: Gujarat's Hope, (no author, no date), Sardar Sarovar Narmada Nigam Limited, Gandhinagar.

Sardar Sarovar Narmada Project (leaflet), (1988), Sardar Sarovar Narmada Nigam Limited, Gandhinagar, October.

Sardar Sarovar Project: The Lifeline of Gujarat, (1988), Sardar Sarovar Narmada Nigam Limited, Gandhinagar.

Singh, Digvijay (Chief Minister of Madhya Pradesh), (1994), letter to Prime Minister P.V. Narasimha Rao, Bhopal, 4 March.

State of the Environment: Gujarat, (2001), Gujarat Ecology Commission, Vadodara, March.

Statistical Outline of India (1994-95), (1994), Tata Services Limited, Bombay, October.

'True Face of Medha Patkar and her NBA,' (2000), paid advertisement in the *Indian Express*, Bombay, 10 November, p. 5. (The ad has appeared in other publications too).

World Development Report, 1997: The State in a Changing World, (1997), published for the World Bank by Oxford University Press, New York.

Index